Aspects of Differential Geometry
IV

Synthesis Lectures on Mathematics and Statistics

Editor
Steven G. Krantz, *Washington University, St. Louis*

Statistics is Easy! Second Edition
Dennis Shasha and Manda Wilson
2010

Lectures on Financial Mathematics: Discrete Asset Pricing
Greg Anderson and Alec N. Kercheval
2010

Jordan Canonical Form: Theory and Practice
Steven H. Weintraub
2009

The Geometry of Walker Manifolds
Miguel Brozos-Vázquez, Eduardo García-Río, Peter Gilkey, Stana Nikčević, and Ramón Vázquez-Lorenzo
2009

An Introduction to Multivariable Mathematics
Leon Simon
2008

Jordan Canonical Form: Application to Differential Equations
Steven H. Weintraub
2008

Statistics is Easy!
Dennis Shasha and Manda Wilson
2008

A Gyrovector Space Approach to Hyperbolic Geometry
Abraham Albert Ungar
2008

Aspects of Differential Geometry IV

Esteban Calviño-Louzao, Eduardo García-Río, Peter Gilkey, JeongHyeong Park, and Ramón Vázquez-Lorenzo

ISBN: 978-3-031-01288-4   paperback
ISBN: 978-3-031-02416-0   ebook
ISBN: 978-3-031-00262-5   hardcover

DOI 10.1007/978-3-031-02416-0

A Publication in the Springer series
SYNTHESIS LECTURES ON MATHEMATICS AND STATISTICS

Lecture #26
Series Editor: Steven G. Krantz, *Washington University, St. Louis*
Series ISSN
Print 1938-1743   Electronic 1938-1751

Aspects of Differential Geometry IV

IV

Esteban Calviño-Louzao
University of Santiago de Compostela, Spain

Eduardo García-Río
University of Santiago de Compostela, Spain

Peter Gilkey
University of Oregon

JeongHyeong Park
SungKyunkan University, Korea

Ramón Vázquez-Lorenzo
University of Santiago de Compostela, Spain

SYNTHESIS LECTURES ON MATHEMATICS AND STATISTICS #26

ABSTRACT

Book IV continues the discussion begun in the first three volumes. Although it is aimed at first-year graduate students, it is also intended to serve as a basic reference for people working in affine differential geometry. It also should be accessible to undergraduates interested in affine differential geometry. We are primarily concerned with the study of affine surfaces which are locally homogeneous. We discuss affine gradient Ricci solitons, affine Killing vector fields, and geodesic completeness. Opozda has classified the affine surface geometries which are locally homogeneous; we follow her classification. Up to isomorphism, there are two simply connected Lie groups of dimension 2. The translation group $\mathbb{R}^2$ is Abelian and the $ax + b$ group is non-Abelian. The first chapter presents foundational material. The second chapter deals with Type $\mathcal{A}$ surfaces. These are the left-invariant affine geometries on $\mathbb{R}^2$. Associating to each Type $\mathcal{A}$ surface the space of solutions to the quasi-Einstein equation corresponding to the eigenvalue $\mu = -1$ turns out to be a very powerful technique and plays a central role in our study as it links an analytic invariant with the underlying geometry of the surface. The third chapter deals with Type $\mathcal{B}$ surfaces; these are the left-invariant affine geometries on the $ax + b$ group. These geometries form a very rich family which is only partially understood. The only remaining homogeneous geometry is that of the sphere S^2. The fourth chapter presents relations between the geometry of an affine surface and the geometry of the cotangent bundle equipped with the neutral signature metric of the modified Riemannian extension.

KEYWORDS

affine gradient Ricci solitons, affine Killing vector fields, geodesic completeness, locally homogeneous affine surfaces, locally symmetric affine surfaces, projectively flat, quasi-Einstein equation

*This book is dedicated to
Alison, Carmen, Celia, Fernanda, Hugo, Junmin, Junpyo, Luis,
Manuel, Mateo, Montse, Rosalía, Sara, and Susana.*

Contents

Preface

This four-volume series arose out of work by the authors over a number of years both in teaching various courses and in their research endeavors. For technical reasons, the material is divided into four books and each book is largely self-sufficient. To facilitate cross references between the books, we have numbered the chapters of Book I from 1–3, the chapters of Book II from 4–8, the chapters of Book III from 9–11, and the chapters of the present Book IV on affine surfaces from 12–15. A final book in the series dealing with elliptic operator theory and its applications to Differential Geometry is proposed.

Up to isomorphism, there are two simply connected Lie groups of dimension 2. The translation group $\mathbb{R}^2$ is Abelian and acts on $\mathbb{R}^2$ by translation; the group structure is given by $(a,b) + (a',b') = (a+a',b+b')$. The $ax + b$ group $\mathbb{R}^+ \times \mathbb{R}$ acts on $\mathbb{R}^+ \times \mathbb{R}$ by

$$(x^1, x^2) \rightarrow (ax^1, ax^2 + b) \quad \text{for} \quad a > 0 \quad \text{and} \quad b \in \mathbb{R};$$

the group structure is given by composition and is non-Abelian;

$$(a,b) * (a',b') = (aa', ab' + b).$$

An affine surface $\mathcal{M}$ is a pair (M, ∇) where M is a smooth surface and ∇ is a torsion-free connection on the tangent bundle of M. One says $\mathcal{M} = (M, \nabla)$ is locally homogeneous if given any two points of M, there is the germ of a diffeomorphism mapping one point to the other point which preserves the connection ∇. Opozda [53] showed that any locally homogeneous affine surface geometry is *modeled* on one of the following three geometries:

- **Type $\mathcal{A}$.** $\mathcal{M} = (\mathbb{R}^2, \nabla)$ where ∇ has constant Christoffel symbols $\Gamma_{ij}{}^k = \Gamma_{ji}{}^k$. This geometry is homogeneous; the Type $\mathcal{A}$ connections are the left-invariant connections on the Lie group $\mathbb{R}^2$. An affine surface is modeled on such a geometry if and only if there exists a coordinate atlas so that the Christoffel symbols ${}^\alpha\Gamma_{ij}{}^k = \Gamma_{ij}{}^k$ are constant in each chart of the atlas.

- **Type $\mathcal{B}$.** $\mathcal{M} = (\mathbb{R}^+ \times \mathbb{R}, \nabla)$ where ∇ has Christoffel symbols $\Gamma_{ij}{}^k = (x^1)^{-1}A_{ij}{}^k$ where $A_{ij}{}^k = A_{ji}{}^k$ is constant. This geometry is homogeneous; the action of the $ax + b$ group sending $(x^1, x^2) \rightarrow (ax^1, ax^2 + b)$ acts transitively on $\mathbb{R}^+ \times \mathbb{R}$. If we identify the $ax + b$ group with $\mathbb{R}^+ \times \mathbb{R}$, then the Type $\mathcal{B}$ connections are the left-invariant connections. An affine surface is modeled on such a geometry if there is a coordinate atlas so the Christoffel symbols ${}^\alpha\Gamma_{ij}{}^k = (x_\alpha^1)^{-1}A_{ij}{}^k$ in each chart of the atlas.

- **Type $\mathcal{C}$.** $\mathcal{M} = (M, \nabla)$ where ∇ is the Levi–Civita connection of a metric of constant non-zero sectional curvature.

This present volume is organized around this observation. There is a non-trivial intersection between the Type $\mathcal{A}$ and the Type $\mathcal{B}$ geometries. There is no geometry which is both Type $\mathcal{A}$ and $\mathcal{C}$. And the only Type $\mathcal{C}$ geometry which is not also Type $\mathcal{B}$ is modeled on the round sphere S^2 in $\mathbb{R}^3$. Chapter 12 of the book deals with preliminary material. We introduce the basics of affine geometry, discuss the affine quasi-Einstein equation, and establish its basic properties. We discuss affine gradient Ricci solitons and other preliminary matters. For surface geometries, the Ricci tensor

$$\rho(x, y) := \mathrm{Tr}\{z \to R(z, x)y\}$$

carries the geometry; an affine connection on a surface is flat if and only if $\rho = 0$.

Chapter 13 is devoted to a discussion of the geometry of Type $\mathcal{A}$ surfaces. Any Type $\mathcal{A}$ surface is strongly projectively flat. The solution space to the quasi-Einstein equation for the critical eigenvalue $\mu = -1$ will play a central role in our discussion as it is a complete invariant of strongly projectively flat surfaces. By identifying the Christoffel symbols $\{\Gamma_{ij}{}^k\}$ with a point of $\mathbb{R}^6$, we parameterize such surfaces. The Ricci tensor of any Type $\mathcal{A}$ surface is symmetric and any such surface is strongly projectively flat. The set of flat Type $\mathcal{A}$ surfaces where Γ does not vanish identically is a smooth 4-dimensional manifold which may be identified with a $\mathbb{Z}_2$ quotient of $S^1 \times S^2 \times \mathbb{R}$. The set of Type $\mathcal{A}$ surfaces where the Ricci tensor has rank 1 and is positive or negative semi-definite is a 5-dimensional manifold which may be identified with $S^1 \times S^1 \times \mathbb{R}^3$. It is natural to identify Type $\mathcal{A}$ geometries which differ by a change of coordinates or, equivalently, by the action of the general linear group $\mathrm{GL}(2, \mathbb{R})$. The resulting moduli spaces of flat Type $\mathcal{A}$ surfaces, of Type $\mathcal{A}$ surfaces where the Ricci tensor has rank 1, and of Type $\mathcal{A}$ surfaces where the Ricci tensor is non-degenerate and has signature (p, q) is determined quite explicitly. The surfaces which are geodesically complete are described up to linear equivalence. We discuss affine Killing vector fields and affine gradient Ricci solitons for such geometries.

In Chapter 14, we present an analogous discussion for the Type $\mathcal{B}$ surfaces. These surfaces are, in general, not strongly projectively flat, and thus the solution space to the quasi-Einstein equation is of less utility. The structure group here is the $ax + b$ group rather than the general linear group where the action this time is $(x^1, x^2) \to (x^1, bx^1 + ax^2)$. Let $\ker_{\mathcal{B}}\{\rho\} - \Gamma_0$ be the the space of flat connections other than the trivial connection where all the Christoffel symbols vanish and let $\ker_{\mathcal{B}}\{\rho_s\} - \ker_{\mathcal{B}}\{\rho\}$ be the space of all connections where the Ricci tensor is purely alternating but does not vanish identically. In contrast to the situation for Type $\mathcal{A}$ geometries, these two spaces are not smooth. The set $\ker_{\mathcal{B}}\{\rho\} - \Gamma_0$ (resp. $\ker_{\mathcal{B}}\{\rho_s\} - \ker_{\mathcal{B}}\{\rho\}$) is an immersed 3-dimensional (resp. 2-dimensional) manifold with transversal intersections. We also discuss affine Killing vector fields and affine gradient Ricci solitons in this context. We determine the locally symmetric Type $\mathcal{B}$ surfaces.

In Chapter 15, we present some applications of affine surface theory. If $\mathcal{M} = (M, \nabla)$ is an affine surface, the modified Riemannian extension gives rise to a neutral signature metric $g_{\nabla, \Phi, T, S, X}$ on the cotangent bundle of M where X is a tangent vector field on M, where Φ is a symmetric 2-tensor on M, and where T and S are endomorphisms of the tangent bundle of M.

There is an intimate relation between the geometry of the affine surface $\mathcal{M}$ and the geometry of $\mathcal{N} := (T^*M, g_{\nabla,\Phi,T,S,X})$. We relate solutions to the affine quasi-Einstein equation on $\mathcal{M}$ and the Riemannian quasi-Einstein equation. We also construct Bach flat signature $(2,2)$ metrics using the Riemannian extension and construct vanishing scalar invariant (VSI) manifolds.

Esteban Calviño-Louzao, Eduardo García-Río, Peter Gilkey, JeongHyeong Park, and Ramón Vázquez-Lorenzo
April 2019

Acknowledgments

We report in this book, among other matters, on the research of the authors and our coauthors and would like, in particular, to note the contributions of D. D'Ascanio, M. Brozos-Vázquez, S. Gavino-Fernández, I. Gutiérrez-Rodríguez, D. Kupeli, S. Nikčević, P. Pisani, X. Valle-Regueiro, and M. E. Vázquez-Abal to this research.

D. D'Ascanio   M. Brozos-Vázquez   S. Gavino-Fernández   I. Gutiérrez-Rodríguez

S. Nikčević   P. Pisani   X. Valle-Regueiro   M. E. Vázquez-Abal

The research of the authors was supported by the Basic Science Research Program through the National Research Foundation of Korea (NRF) funded by the Ministry of Education (NRF-2016R1D1A1B03930449) and by Projects EM2014/009, MTM2013-41335-P, and MTM2016-75897-P (AEI/FEDER, UE). The assistance of Ekaterina Puffini of the Krill Institute of Technology has been invaluable.

Esteban Calviño-Louzao, Eduardo García-Río, Peter Gilkey, JeongHyeong Park, and Ramón Vázquez-Lorenzo
April 2019

CHAPTER 12

An Introduction to Affine Geometry

In Section 12.1, we summarize some basic definitions and facts. We discuss the curvature operator, the Ricci tensor, affine structures, and geodesics. We give a brief introduction to the theory of affine symmetric spaces and projective equivalence. We recall the definition of the Lie derivative and the Lie algebra of affine Killing vector fields. We define the Hessian, affine gradient Yamabe solitons, the affine quasi-Einstein operator, and affine gradient Ricci solitons. In Section 12.2, we present the classification of Wong of recurrent surfaces. In Section 12.3, we discuss the affine quasi-Einstein equation in the context of affine geometry. We prove the basic facts that we shall be using; strong projective equivalence plays an important role in our discussion. We exhibit some inhomogeneous examples. In Section 12.4 we discuss results of Opozda concerning locally affine homogeneous surfaces. In Section 12.5, we show that any locally affine homogeneous surface has a natural real analytic structure. We also discuss the structure of various spaces of smooth functions and other natural tensors as modules over the two 2-dimensional Lie algebras.

12.1 BASIC DEFINITIONS

In this section, we introduce the basic concepts we will be examining throughout the book. Section 12.1.1 discusses curvature, Section 12.1.2 introduces the theory of affine Killing vector fields, Section 12.1.3 treats geodesics, Section 12.1.4 treats geodesic completeness, and Section 12.1.5 defines affine maps, symmetric spaces, and models. Projective equivalence is introduced in Section 12.1.6. The Hessian and the affine quasi-Einstein operator (see Section 12.1.7) will play a central role in our development. Gradient Ricci and Yamabe solitons (see Section 12.1.8) and affine gradient Ricci and Yamabe solitons (see Section 12.1.9) will be examined in more detail in subsequent chapters. Classical results in ellipticity are given in Section 12.1.10.

Let $\vec{x} = (x^1, \ldots, x^m)$ be a system of local coordinates on a smooth manifold M. We adopt the *Einstein convention* and sum over repeated indices henceforth. Let $\partial_{x^i} := \frac{\partial}{\partial x^i}$ and let $\theta^{(a_1,\ldots,a_m)} := (\partial_{x^1})^{a_1} \ldots (\partial_{x^m})^{a_m} \theta$. Let ∇ be a connection on the tangent bundle of M. Expand $\nabla_{\partial_{x^i}} \partial_{x^j} = \Gamma_{ij}{}^k \partial_{x^k}$ where $\Gamma = (\Gamma_{ij}{}^k)$ are the *Christoffel symbols* of the connection. We say that ∇ is *torsion-free* and that M is an *affine manifold* if $\nabla_X Y - \nabla_Y X - [X, Y] = 0$ or, equivalently, if $\Gamma_{ij}{}^k = \Gamma_{ji}{}^k$. The importance of the torsion-free condition lies in the observation that ∇ is

2 12. AN INTRODUCTION TO AFFINE GEOMETRY

torsion-free if and only if for every point P of M, there exist coordinates centered at P so that $\Gamma_{ij}{}^k(P) = 0$ (see Lemma 3.5 of Book I). We shall assume ∇ is torsion-free in this volume and refer to Arias-Marco and Kowalski [2], Gilkey [31, 32], and forthcoming work by D'Ascanio, Gilkey, and Pisani [23] for some results concerning the geometry of surfaces if the connection is permitted to have torsion.

12.1.1 CURVATURE. The *curvature operator* and the *Ricci tensor* are defined, respectively, by setting

$$R(X, Y) := \nabla_X \nabla_Y - \nabla_Y \nabla_X - \nabla_{[X,Y]},$$
$$\rho(X, Y) := \mathrm{Tr}\{Z \to R(Z, X)Y\}.$$

The components of these tensors may be expressed locally in the form:

$$
\begin{aligned}
R_{ijk}{}^\ell &= \partial_{x^i} \Gamma_{jk}{}^\ell - \partial_{x^j} \Gamma_{ik}{}^\ell + \Gamma_{in}{}^\ell \Gamma_{jk}{}^n - \Gamma_{jn}{}^\ell \Gamma_{ik}{}^n, \\
\rho_{jk} &= \partial_{x^i} \Gamma_{jk}{}^i - \partial_{x^j} \Gamma_{ik}{}^i + \Gamma_{in}{}^i \Gamma_{jk}{}^n - \Gamma_{jn}{}^i \Gamma_{ik}{}^n.
\end{aligned}
\tag{12.1.a}
$$

In the affine setting, unlike in the pseudo-Riemannian context, the Ricci tensor need not be symmetric. Thus we introduce the *symmetric Ricci tensor* ρ_s and the *alternating Ricci tensor* ρ_a by defining:

$$\rho_s(X, Y) := \tfrac{1}{2}\{\rho(X, Y) + \rho(Y, X)\} \quad \text{and} \quad \rho_a(X, Y) := \tfrac{1}{2}\{\rho(X, Y) - \rho(Y, X)\}. \tag{12.1.b}$$

The following result shows that Ricci tensor carries the geometry for affine surfaces.

Lemma 12.1 *If $\mathcal{M} = (M, \nabla)$ is an affine surface, then $\mathcal{M}$ is flat if and only if $\rho = 0$.*

Proof. Let $\mathcal{M}$ be an affine surface. We have

$$\rho_{11} = R_{211}{}^2, \quad \rho_{12} = R_{212}{}^2, \quad \rho_{21} = R_{121}{}^1, \quad \rho_{22} = R_{122}{}^1. \tag{12.1.c}$$

Thus the Ricci tensor determines the curvature operator; $\rho = 0$ if and only if $R = 0$ in dimension 2. $\qquad\square$

12.1.2 AFFINE KILLING VECTOR FIELDS. If X is a smooth vector field on M, let Φ_t^X be the local flow defined by X. The *Lie derivative* $\mathcal{L}_X(\Xi)$ is defined by

$$\mathcal{L}_X(\Xi) := \lim_{t \to 0} \frac{\Phi_{-t}^X \circ \Xi \circ \Phi_t^X - \Xi}{t}.$$

If f is a function, then $\mathcal{L}_X(f) = X(f)$. If Y is a vector field, then $\mathcal{L}_X(Y) = [X, Y]$. If ω is a 1-form, then $\mathcal{L}_X \omega$ is characterized by the identity $\langle \mathcal{L}_X(Y), \omega \rangle + \langle Y, \mathcal{L}_X \omega \rangle = X\langle Y, \omega \rangle$. We refer to Kobayashi and Nomizu [39, Chapter VI] for the proof of the following result.

Lemma 12.2   *Let $\mathcal{M} = (M, \nabla)$ be an affine manifold of dimension m, possibly with torsion.*

1. *The following three conditions are equivalent and if any is satisfied, then X is said to be an* affine Killing vector field*:*

   (a) $(\Phi_t^X)_* \circ \nabla = \nabla \circ (\Phi_t^X)_*$ on the appropriate domain.

   (b) The Lie derivative $\mathcal{L}_X(\nabla)$ of ∇ vanishes.

   (c) $[X, \nabla_Y Z] - \nabla_Y [X, Z] - \nabla_{[X,Y]} Z = 0$ for all $Y, Z \in C^\infty(TM)$.

2. *Let $\mathfrak{K}(\mathcal{M})$ be the set of affine Killing vector fields. The Lie bracket gives $\mathfrak{K}(\mathcal{M})$ the structure of a real Lie algebra.*

The equation $[X, \nabla_Y Z] - \nabla_Y [X, Z] - \nabla_{[X,Y]} Z = 0$ is tensorial in $\{Y, Z\}$ so we may take $Y = \partial_{x^i}$ and $Z = \partial_{x^j}$. Expand $X = a^n \partial_{x^n}$. We compute:

$$[X, \nabla_Y Z] = [a^\ell \partial_{x^\ell}, \Gamma_{ij}{}^n \partial_{x^n}] = a^\ell \partial_{x^\ell} \Gamma_{ij}{}^n \partial_{x^n} - \Gamma_{ij}{}^n \partial_{x^n} a^\ell \partial_{x^\ell}$$

$$= \{a^\ell \partial_{x^\ell} \Gamma_{ij}{}^k - \Gamma_{ij}{}^\ell \partial_{x^\ell} a^k\} \partial_{x^k},$$

$$-\nabla_Y [X, Z] = -\nabla_{\partial_{x^i}} [a^\ell \partial_{x^\ell}, \partial_{x^j}] = \nabla_{\partial_{x^i}} (\partial_{x^j} a^\ell \partial_{x^\ell})$$

$$= \partial_{x^i} \partial_{x^j} a^k \partial_{x^k} + \Gamma_{i\ell}{}^k \partial_{x^j} a^\ell \partial_{x^k},$$

$$-\nabla_{[X,Y]} Z = -\nabla_{[a^\ell \partial_{x^\ell}, \partial_{x^i}]} \partial_{x^j} = \Gamma_{\ell j}{}^k \partial_{x^i} a^\ell \partial_{x^k}.$$

We now obtain m^3 affine Killing equations for $1 \le i, j, k \le m$ from Assertion 1-c:

$$K_{ij}{}^k : \quad 0 = \frac{\partial^2 a^k}{\partial x^i \partial x^j} + \sum_\ell \left\{ a^\ell \frac{\partial \Gamma_{ij}{}^k}{\partial x^\ell} - \Gamma_{ij}{}^\ell \frac{\partial a^k}{\partial x^\ell} + \Gamma_{i\ell}{}^k \frac{\partial a^\ell}{\partial x^j} + \Gamma_{\ell j}{}^k \frac{\partial a^\ell}{\partial x^i} \right\}. \qquad (12.1.d)$$

Let $\mathcal{M}$ be an affine manifold. We say that $\mathcal{M}$ is *affine Killing complete* if every affine Killing vector field of $\mathcal{M}$ is a complete vector field, i.e., the integral curves exist for all time. We examine this notion further in Section 13.2.7 and in Section 14.2.5. Let $\mathrm{Aff}(\mathcal{M})$ be the group of all affine diffeomorphisms of an affine manifold $\mathcal{M}$.

Lemma 12.3   *Let $\mathcal{M}$ be a connected affine manifold $\mathcal{M}$.*

1. $\mathrm{Aff}(\mathcal{M})$ *is a Lie group.*

2. *The Lie algebra $\mathfrak{a}(\mathcal{M})$ of $\mathrm{Aff}(\mathcal{M})$ is the space of complete affine Killing vector fields.*

3. $\mathcal{M}$ *is affine Killing complete if and only if $\mathfrak{a}(\mathcal{M}) = \mathfrak{K}(\mathcal{M})$.*

Proof. We sketch the proof briefly and refer to Kobayashi and Nomizu [39] for a more complete explanation. Let $\mathcal{F}(\mathcal{M})$ be the frame bundle of the tangent bundle; this is a principal $\mathrm{GL}(m, \mathbb{R})$ bundle over M. Fix a positive definite metric on $\mathbb{R}^m$ and a left-invariant Riemannian metric on $\mathrm{GL}(m, \mathbb{R})$, i.e., a positive definite inner product on the Lie algebra $\mathfrak{gl}(m, \mathbb{R})$ of $\mathrm{GL}(m, \mathbb{R})$. The connection ∇ (which may be permitted to have torsion) defines a canonical splitting of the tangent bundle of $\mathcal{F}(M)$ into vertical and horizontal subspaces. At a point F of $\mathcal{F}(M)$, we may use the frame to identify the horizontal subspace with $\mathbb{R}^m$ and the vertical subspace with $\mathfrak{gl}(m)$. We use the metrics on $\mathbb{R}^m$ and $\mathfrak{gl}(m)$ to define a Riemannian metric on $\mathcal{F}(M)$. If Φ is a (local) affine map, then the natural lift of Φ to the frame bundle is a (local) isometry of $\mathcal{F}(M)$. This defines a natural embedding of $\mathrm{Aff}(M)$ as a closed subgroup of the group of isometries of $\mathcal{F}(M)$. The group of isometries of a Riemannian manifold is a Lie group (see Theorem 7.6 of Book II). A closed subgroup of a Lie group is again a Lie group (see Theorem 6.10 of Book II). It now follows that $\mathrm{Aff}(M)$ is a Lie group. The remaining two Assertions are now immediate. $\square$

The following is a useful observation that will play an important role in the classification of locally homogeneous surfaces in Section 12.4.

Lemma 12.4   *Let $M = (M, \nabla)$ be an affine surface, possibly with torsion. Let X be an affine Killing vector field with $X(P) \neq 0$ at a point P of M. We can choose local coordinates centered at P so that*

$$X = \partial_{x^2}, \quad \Gamma_{ij}{}^k(x^1, x^2) = \Gamma_{ij}{}^k(x^1), \quad \Gamma_{11}{}^1(x^1) = 0, \quad \Gamma_{11}{}^2(x^1) = 0.$$

Proof. Choose initial coordinates (y^1, y^2) centered at P so that $X = \partial_{y^2}$. Since X is an affine Killing vector field, the Christoffel symbols do not depend on y^2, i.e., we may express $\Gamma_{ij}{}^k(y^1, y^2) = \Gamma_{ij}{}^k(y^1)$; the map $(y^1, y^2) \to (y^1, y^2 + t)$ is then an affine map. Let $\sigma(s) = (y^1(s), y^2(s))$ be an affine geodesic with $\sigma(0) = (0, 0)$ and with $\dot{\sigma}(0)$ and $X(0)$ linearly independent. Let $T(x^1, x^2) := (y^1(x^1), y^2(x^1) + x^2)$ define new coordinates with $\partial_{x^2} = \partial_{y^2}$. Since the curves $x^1 \to T(x^1, x^2)$ are affine geodesics for x^2 fixed and since ∂_{x^2} is an affine Killing vector field, the normalizations of the result hold. $\square$

We present a technical lemma, which we shall need subsequently in Section 12.4, illustrating how the affine Killing equations are used. Let

$$\mathfrak{K}_{\mathbb{C}}(M) := \mathfrak{K}(M) \otimes_{\mathbb{R}} \mathbb{C}$$

be the algebra of complex affine Killing vector fields.

Lemma 12.5   *Let $M = (M, \nabla)$ be an affine surface, possibly with torsion. Normalize the coordinates as in Lemma 12.4 to assume*

$$\Gamma_{ij}{}^k(x^1, x^2) = \Gamma_{ij}{}^k(x^1), \quad \Gamma_{11}{}^1(x^1) = 0, \quad \Gamma_{11}{}^2(x^1) = 0.$$

1. If $0 \neq v(x^1)\partial_{x^2} \in \mathfrak{K}(M)$, with $v'(x^1) \neq 0$, then

$$\begin{aligned} \Gamma_{11}{}^1 = 0, \quad \Gamma_{11}{}^2 = 0, \quad \Gamma_{12}{}^1 = 0, \quad \Gamma_{21}{}^1 = 0, \\ \Gamma_{22}{}^1 = 0, \quad \Gamma_{22}{}^2 = 0, \quad (\Gamma_{12}{}^2 + \Gamma_{21}{}^2)v' + v'' = 0. \end{aligned} \tag{12.1.e}$$

If $\{u_1(x^1)\cos(x^2) + u_2(x^1)\sin(x^2)\}\partial_{x^1} + w(x^1, x^2)\partial_{x^2}$ belongs to $\mathfrak{K}(M)$ and if Γ satisfies the relations of Equation (12.1.e), then $u_1 = u_2 = 0$.

2. *Assume* $0 \neq \alpha \in \mathbb{C}$. *If* $0 \neq e^{\alpha x^2} v(x^1)\partial_{x^2} \in \mathfrak{K}_{\mathbb{C}}(\mathcal{M})$, *then*

$$\Gamma_{11}{}^1 = 0, \quad \Gamma_{11}{}^2 = 0, \quad \Gamma_{12}{}^1 = 0, \quad \Gamma_{21}{}^1 = 0,$$
$$\Gamma_{22}{}^1 = 0, \quad \Gamma_{22}{}^2 = -\alpha. \tag{12.1.f}$$

If $X := e^{\alpha x^2} u(x^1)\partial_{x^1} + w(x^1, x^2)\partial_{x^2}$ *belongs to* $\mathfrak{K}_{\mathbb{C}}(\mathcal{M})$ *and if* Γ *satisfies the relations of Equation (12.1.f), then* $u(x^1) = 0$.

Proof. The relations of Equation (12.1.e) follow from the affine Killing equations for $v(x^1)\partial_{x^2}$:

$$K_{11}{}^1 : \ 0 = (\Gamma_{12}{}^1 + \Gamma_{21}{}^1)v', \qquad K_{12}{}^1 : \ 0 = \Gamma_{22}{}^1 v',$$
$$K_{12}{}^2 : \ 0 = (\Gamma_{22}{}^2 - \Gamma_{12}{}^1)v', \qquad K_{21}{}^2 : \ 0 = (-\Gamma_{21}{}^1 + \Gamma_{22}{}^2)v',$$
$$K_{11}{}^2 : \ 0 = (\Gamma_{12}{}^2 + \Gamma_{21}{}^2)v' + v''.$$

Suppose that $\{u_1(x^1)\cos(x^2) + u_2(x^1)\sin(x^2)\}\partial_{x^1} + w(x^1, x^2)\partial_{x^2}$ is an affine Killing vector field and the relations of Equation (12.1.e) hold. We then have

$$K_{22}{}^1 : \ 0 = -u_1(x^1)\cos(x^2) - u_2(x^1)\sin(x^2).$$

Consequently, $u_1 = u_2 = 0$. This proves Assertion 1

Assume $0 \neq e^{\alpha x^2} v(x^1)\partial_{x^2} \in \mathfrak{K}_{\mathbb{C}}(\mathcal{M})$ for $0 \neq \alpha \in \mathbb{C}$. We have $\Gamma_{11}{}^1 = 0$, $\Gamma_{11}{}^2 = 0$, $\Gamma_{ij}{}^k = \Gamma_{ij}{}^k(x^1)$, and $\partial_{x^2} \in \mathfrak{K}(\mathcal{M})$. We evaluate all affine Killing equations at $x^2 = 0$ to eliminate the exponential. The relations of Equation (12.1.f) follow from the affine Killing equations

$$K_{22}{}^1 : \ 0 = 2\alpha\Gamma_{22}{}^1 v, \quad \text{so} \quad \Gamma_{22}{}^1 = 0,$$
$$K_{12}{}^1 : \ 0 = \alpha\Gamma_{12}{}^1 v + \Gamma_{22}{}^1 v', \quad \text{so} \quad \Gamma_{12}{}^1 = 0,$$
$$K_{21}{}^1 : \ 0 = \alpha\Gamma_{21}{}^1 v + \Gamma_{22}{}^1 v', \quad \text{so} \quad \Gamma_{21}{}^1 = 0,$$
$$K_{22}{}^2 : \ 0 = \alpha v(\alpha + \Gamma_{22}{}^2) - \Gamma_{22}{}^1 v', \quad \text{so} \quad \Gamma_{22}{}^2 = -\alpha.$$

Assume that $e^{\alpha x^2} u(x^1)\partial_{x^1} + w(x^1, x^2)\partial_{x^2} \in \mathfrak{K}_{\mathbb{C}}(\mathcal{M})$ and that the relations of Equation (12.1.f) hold. The associated affine Killing equation $K_{22}{}^1 : \ 0 = 2\alpha^2 u(x^1)$ implies that $u(x^1) = 0$ as desired and establishes Assertion 2. $\square$

12.1.3 GEODESICS.
We say that a curve γ in an affine manifold is an *affine geodesic* if $\ddot{\gamma} = 0$ or, in other words, $\nabla_{\dot{\gamma}}\dot{\gamma} = 0$. Express $\gamma(t) = (x^1(t), \dots, x^m(t))$ in a system of local coordinates. Then γ is an affine geodesic if and only if the *geodesic equation* is satisfied:

$$\ddot{x}^i(t) + \Gamma_{jk}{}^i(x(t))\dot{x}^j(t)\dot{x}^k(t) = 0.$$

Fix a point P of M and a tangent vector $\xi \in T_P M$. The Fundamental Theorem of Ordinary Differential Equations shows that the geodesic equation has a unique solution with initial conditions $x(0) = P$ and $\dot{x}(0) = \xi$ for $|t| < \varepsilon$. We say that $\mathcal{M}$ is *geodesically complete* if $\varepsilon(P, \xi) = \infty$ for every (P, ξ), or, in other words, if every affine geodesic extends for infinite time. We say that a submanifold N of M is *totally geodesic* if any affine geodesic in M which is tangent to N at a single point is entirely contained in N. We will discuss these notions further in the context of affine homogeneous surfaces. The *exponential map* $\exp_P$ (see Section 3.4.1 of Book I in the Riemannian setting) is a local diffeomorphism from a neighborhood of the origin in $T_P M$ to a neighborhood of P in M which is characterized by the fact that the curves $t \to \exp_P(t\xi)$ are affine geodesics starting at P with initial direction ξ. A bit of caution is needed as there are significant differences between the affine setting and the Riemannian setting.

Suppose given two affine manifolds $\mathcal{M}_1 = (M_1, \nabla_1)$ and $\mathcal{M}_2 = (M_2, \nabla_2)$. We say that a diffeomorphism Ψ from M_1 to M_2 is an *affine map* if Ψ intertwines the two connections, i.e., if $\Psi^* \nabla_2 = \nabla_1$.

Lemma 12.6 *If Φ is an affine diffeomorphism of an affine manifold $\mathcal{M}$, then the fixed point set of Φ is the union of smooth totally geodesic submanifolds.*

Proof. Let F be the fixed point set of Φ. If $P \in F$, let $V_0(P) := \{\xi \in T_P M : \Phi_* \xi = \xi\}$. Choose a small neighborhood $\mathcal{U}$ of the origin in $T_P M$ so that $\exp_P$ is a diffeomorphism from $\mathcal{U}$ to a neighborhood $\mathcal{O}$ of P in M and so that there is a unique affine geodesic joining any two points of $\mathcal{O}$. If $\xi \in T_P M$, then $\Phi_* \xi = \xi$ if and only if Φ fixes the affine geodesic through P with initial direction ξ. Furthermore, if $Q \in F \cap \mathcal{O}$, then there is a unique affine geodesic σ between P and Q lying in $\mathcal{O}$. Consequently, if ε is sufficiently small and if $Q \in \exp_P(B_\varepsilon(0))$, then $\Phi\sigma = \sigma$. This shows that $\exp_P(V_0 \cap B_\varepsilon(0)) = \mathcal{O} \cap F$. $\qquad\square$

12.1.4 GEODESIC COMPLETENESS.
If $\mathcal{M}$ is an affine manifold, we say that $\mathcal{M}$ is *geodesically complete* if every geodesic extends for infinite time; if $\mathcal{M}$ is not geodesically complete, we say $\mathcal{M}$ is *geodesically incomplete*.

Let $\mathcal{M}$ be the circle with the usual periodic parameter θ defined modulo 2π and with the connection given by $\Gamma_{11}{}^1 = 1$. The geodesic equation becomes $\ddot{\theta} + \dot{\theta}\dot{\theta} = 0$. Set $\theta(t) = \log(t)$ mod 2π for $t \in (0, \infty)$. Then $\dot{\theta} = t^{-1}$ and $\ddot{\theta} = -t^{-2}$. Thus this satisfies the geodesic equation and the maximal parameter range is $t \in (0, \infty)$. This provides an example of a compact affine geometry which is geodesically incomplete.

Let $\mathcal{M}_1$ be $S^1 \times [1, 4]$ with coordinates (θ, r) where θ is defined modulo 2π, where r belongs to the interval $[1, 4]$, and where $\Gamma_{11}{}^1 = 1$ is the only non-zero Christoffel symbol. Use polar coordinates to embed $S^1 \times [1, 4]$ as an annulus in the plane and use a partition of unity to extend the affine structure to the plane keeping the original geometry on $S^1 \times [1, 4]$ to obtain a flat geometry near infinity and near the origin. Again, we have an incomplete affine geodesic; this affine geodesic remains bounded, but the velocity increases without bound. It simply goes around the origin faster and faster on the circle of radius 2.

Sometimes there is an apparent singularity which can be removed. Let the non-zero Christoffel symbols of $\mathcal{M}$ be $\Gamma_{11}{}^1 = 1$, $\Gamma_{12}{}^2 = 1$, and $\Gamma_{22}{}^1 = -1$. Let $\arctan(\cdot)$ map $\mathbb{R}$ to

$(-\frac{\pi}{2}, \frac{\pi}{2})$. One verifies that the two curves

$$\sigma_1(t) := (\log(1+t), 0),$$

$$\sigma_2(t) := \left(\frac{1}{2}\log((1+t)^2 + t^2), \arctan\left(\frac{t}{1+t}\right)\right) \quad \text{for} \quad t \in (-1, \infty)$$

are affine geodesics. The first curve has a genuine singularity at $t = -1$. However, the apparent singularity at $t = -1$ in the second curve can be removed by taking a different branch of the arctangent function and setting

$$\tilde{\sigma}_2(t) := \left\{ \begin{array}{ll} \left(\frac{1}{2}\log((1+t)^2 + t^2), \arctan\left(\frac{t}{1+t}\right)\right) & \text{for } t \in (-1, \infty) \\ (0, -\frac{\pi}{2}) & \text{for } t = -1 \\ \left(\frac{1}{2}\log((1+t)^2 + t^2), \arctan\left(\frac{t}{1+t}\right) - \pi\right) & \text{for } t \in (-\infty, -1) \end{array} \right\}.$$

One has the following picture where the upper curve is σ_2 and the lower curve is $\tilde{\sigma}_2$.

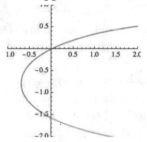

The crucial point in this example is, of course, that $\lim_{t \to -1+} \sigma_2(t)$ exists. We have:

Lemma 12.7 *Let P be a point of an affine manifold $\mathcal{M}$. Let $\sigma : [0, T) \to \mathcal{M}$ be an affine geodesic. Suppose $\lim_{t \to T} \sigma(t) = P$ exists. Then there exists $\varepsilon > 0$ so that σ can be extended to the parameter range $[0, T + \varepsilon)$ as an affine geodesic.*

Proof. Put a positive definite inner product $\langle \cdot, \cdot \rangle$ on $T_P\mathcal{M}$ to act as a reference metric. Let B_r be the ball of radius r about the origin in $T_P\mathcal{M}$. Since the exponential map is a local diffeomorphism, we can use $\exp_P$ to identify B_ε with a neighborhood of P in $\mathcal{M}$ for some small ε. We use this identification to define a flat Riemannian metric near P on $\mathcal{M}$ so that $\exp_P$ is an isometry from B_ε to $\mathcal{M}$. Let $d(\cdot, \cdot)$ be the associated distance function on $\mathcal{M}$. Let $B_r(P) := \exp_P(B_r) = \{Q : d(P, Q) \le r\}$ for $r \le \varepsilon$. Choose linear coordinates on $T_P\mathcal{M}$ to put coordinates on $B_\varepsilon(P)$. This identifies $T_Q\mathcal{M}$ with $T_P\mathcal{M}$ and extends $\langle \cdot, \cdot \rangle$ to $T_Q\mathcal{M}$. Compactness shows that there exists $0 < \tau < \frac{1}{2}\varepsilon$ so that if $Q \in B_{\frac{\varepsilon}{2}}(P)$ and if $\xi \in T_Q\mathcal{M}$ satisfies $\|\xi\| = 1$, then the affine geodesic $\sigma_{Q,\xi}(t) := \exp_Q(t\xi)$ exists for $t \in [0, \tau]$ and belongs to $B_\varepsilon(P)$. By continuity, we can choose $0 < \delta < \frac{1}{4}\tau$ so that if $Q \in B_\delta(P)$ and $\|\xi\| = 1$, then $d(\sigma_{Q,\xi}(\tau), \sigma_{P,\xi}(\tau)) < \frac{\tau}{2}$. Since $d(P, \sigma_{P,\xi}(\tau)) = \tau$, this implies $d(P, \sigma_{Q,\xi}(\tau)) \ge \frac{1}{2}\tau$. We conclude from these estimates that any non-trivial affine geodesic which begins in $B_\delta(P)$ continues to exist at least until it exits from $B_{\frac{1}{2}\tau}(P)$ and that it does in fact exit from $B_{\frac{1}{2}\tau}(P)$.

We assumed $\lim_{t \to T} \sigma(t) = P$. Choose $T_0 < T$ so $\sigma(T_0, T) \subset B_\delta(P)$. Then σ continues to exist until σ exits from $B_{\frac{1}{2}\tau}(P)$ and in particular extends to an affine geodesic defined on $(T_0, T + \varepsilon)$ for some ε.   □

Suppose that M is *locally affine homogeneous*. We say that another affine manifold $\tilde{M}$ is *modeled* on M if every point $\tilde{P}$ of $\tilde{M}$ admits a neighborhood which is affine diffeomorphic to some small open subset of M; the precise open subset being irrelevant as M was assumed locally affine homogeneous. We say that M is *essentially geodesically incomplete* if there is no locally affine homogeneous surface $\tilde{M}$ which is modeled on M which is geodesically complete. The following provides useful criteria.

Lemma 12.8   *Let M be an affine homogeneous surface.*

1. *Suppose there exists an affine geodesic σ in M defined for $t \in (0, T)$ for $T < \infty$ such that $\lim_{t \to T} \rho(\dot{\sigma}(t), \dot{\sigma}(t)) = \pm\infty$. Then M is essentially geodesically incomplete.*

2. *Suppose there exists an affine Killing vector field X on M and an affine geodesic σ in M defined for $t \in (0, T)$ for $T < \infty$ such that $\lim_{t \to T} \rho(X(t), \dot{\sigma}(t)) = \pm\infty$. Then M is essentially geodesically incomplete.*

Proof. Let $f_\sigma(t) := \rho(\dot{\sigma}(t), \dot{\sigma}(t))$ such that $\lim_{t \to T} f_\sigma(t) = \pm\infty$ for some finite value of T. This clearly shows M is geodesically incomplete. Suppose, however, that there exists a geodesically complete geometry $\tilde{M}$ which is modeled on M. We argue for a contradiction. We will show in Theorem 12.52 that the coordinate atlas for $\tilde{M}$ is real analytic. Copy a small piece of σ into $\tilde{M}$ to define an affine geodesic $\tilde{\sigma}$. The function $\tilde{f}(t) := \tilde{\rho}(\dot{\tilde{\sigma}}, \dot{\tilde{\sigma}})$ would then agree with $f(t)$ on some parameter range. Since we are in the real analytic setting, we then have $\tilde{f}$ agrees with f on $(0, T)$ and thus $\lim_{t \to T} \tilde{f}(t) = \pm\infty$. This contradicts the assumption that $\tilde{M}$ was geodesically complete. Consequently, M is essentially geodesically incomplete; this contradiction establishes Assertion 1. The proof of Assertion 2 is the same since, as we shall show in Lemma 12.14, any affine Killing vector field is real analytic.   □

12.1.5 AFFINE MAPS, SYMMETRIC SPACES, AND MODELS. An affine manifold M is said to be *affine homogeneous* if given any two points P and $\tilde{P}$ of M, there is an affine map from M to M taking P to $\tilde{P}$. Correspondingly, M is said to be *locally affine homogeneous* if Ψ is only defined from some neighborhood of P to some neighborhood of $\tilde{P}$. Suppose that M_0 is affine homogeneous. We say that M is *modeled* on M_0 if there is an open neighborhood $\mathcal{O}_P$ of any point P of M and an affine diffeomorphism Φ from $\mathcal{O}_P$ to some open set of M_0. We say that M is *locally affine symmetric* if $\nabla R = 0$.

Theorem 12.9   *Let M be a connected locally affine symmetric manifold.*

1. *M is locally affine homogeneous.*

2. *If ρ_s has maximal rank, then ∇ is the Levi-Civita connection of the locally affine symmetric pseudo-Riemannian manifold (M, ρ_s).*

3. *If M is 2-dimensional, then the Ricci tensor of M is symmetric.*

Proof. We establish Assertion 1 as follows. The exponential map is a diffeomorphism from an open neighborhood $\mathcal{O}$ of 0 in $T_P M$ to an open neighborhood $\tilde{\mathcal{O}}$ of P in M. We may assume that $-\mathcal{O} = \mathcal{O}$ without loss of generality. The *geodesic symmetry* is defined by setting

$$\sigma_P(Q) := \exp_P(-\exp_P^{-1}(Q)) \text{ for } Q \in \tilde{\mathcal{O}}.$$

Work of Nomizu [47] (see Theorem 17.1) shows that σ_P is an affine map. One can compose geodesic symmetries around various points to show that M is locally affine homogeneous. We refer to Koh [41] for subsequent related work. We also note that if M is locally affine symmetric, then M is k-affine curvature homogeneous for all k and this result follows from the work of Pecastaing [57] on the "Singer number" in a quite general context. Finally, Theorem 12.9 follows from work of Opozda [52] in the real analytic setting.

If (M, g) is a pseudo-Riemannian manifold, then there exists a unique affine connection so $\nabla^g g = 0$ (see Theorem 3.7 of Book I). Suppose ρ_s has rank m. Then (M, ρ_s) is a pseudo-Riemannian manifold. Furthermore, $\nabla = \nabla^g$. Since $\nabla R = 0$, $\nabla \rho_s = 0$ and (M, g) is a locally symmetric pseudo-Riemannian manifold. Assertion 2 follows.

Let R be the curvature operator; $R_{ij}(\partial_{x^k}) = R_{ijk}{}^\ell \partial_{x^\ell}$. We have a dual action on the cotangent bundle given by $R_{ij}(dx^\ell) = -R_{ijk}{}^\ell dx^k$. We extend this action to the whole exterior algebra by the Leibnitz rule. Let $\rho_a = (\rho_{12} - \rho_{21}) dx^1 \otimes dx^2$ be the alternating Ricci tensor defined in Equation (12.1.b). Expand $\nabla^2 \rho_a = \rho_{a;ij}(dx^1 \wedge dx^2) \otimes dx^i \otimes dx^j$, where

$$\rho_{a;ij} = \nabla_{\partial_{x^j}} \nabla_{\partial_{x^i}} \rho_a.$$

The commutator of covariant differentiation is given by curvature. We compute

$$
\begin{aligned}
\rho_{a;21} - \rho_{a;12} &= (\rho_{12} - \rho_{21}) R_{12}(dx^1 \wedge dx^2) \\
&= (\rho_{12} - \rho_{21})(-R_{121}{}^1 - R_{122}{}^2)(dx^1 \wedge dx^2) \\
&= (\rho_{12} - \rho_{21})(-\rho_{21} + \rho_{12})(dx^1 \wedge dx^2).
\end{aligned}
$$

If $\nabla \rho = 0$, then $\nabla \rho_a = 0$ and thus $(\rho_{12} - \rho_{21})^2 = 0$. This proves Assertion 3. We remark that Assertion 3 was known to Opozda [50]; the proof we have given is different from the proof appearing there and is due to D'Ascanio, Gilkey, and Pisani [23].   □

Remark 12.10   If the connection on the tangent bundle is not assumed to be torsion-free, then the situation is very different.

1. Let the non-zero Christoffel symbols of $M = (\mathbb{R}^2, \nabla)$ be given by

$$\Gamma_{12}{}^2 = -\frac{1 - e^{2x^1}}{1 + e^{2x^1}} \quad \text{and} \quad \Gamma_{21}{}^2 = \frac{1 - e^{2x^1}}{1 + e^{2x^1}};$$

then a direct computation shows $\rho = dx^1 \otimes dx^1$ and $\nabla\rho = 0$. The space of affine Killing vector fields (see Section 12.1.2) takes the form span$\{\partial_{x^2}, x^1\partial_{x^2}, x^2\partial_{x^2}\}$ so this geometry is not locally affine homogeneous (although it is *cohomogeneity 1*, i.e., the orbits of the affine group have codimension 1) and thus Assertion 1 fails in this context.

2. Let the non-zero Christoffel symbols be given by

$$\Gamma_{11}{}^1 = -\frac{1}{x^1}, \quad \Gamma_{12}{}^2 = -\frac{1}{x^1}, \quad \Gamma_{21}{}^2 = -\frac{3}{x^1}, \quad \Gamma_{22}{}^1 = \frac{1}{x^1}.$$

This is a Type $\mathcal{B}$ affine surface with torsion (see Definition 12.42). We then have

$$\rho = (x^1)^{-2}\{-3dx^1 \otimes dx^1 - dx^2 \otimes dx^2\} \text{ and } \nabla\rho = 0.$$

Since the connection in question has torsion, it is not the Levi-Civita connection of any pseudo-Riemannian metric. Thus Assertion 2 also fails in this context. Note that in contrast to the previous example, this structure is homogeneous since $x^1\partial_{x^1} + x^2\partial_{x^2}$ and ∂_{x^2} are affine Killing vector fields for this geometry.

3. The proof we gave of Assertion 3 remains valid even if torsion is present; if $\mathcal{M}$ is an affine surface with torsion with $\nabla\rho = 0$, then ρ is symmetric. We refer to D'Ascanio, Gilkey, and Pisani [23] for further details.

12.1.6 PROJECTIVE EQUIVALENCE. Two connections ∇ and $\tilde{\nabla}$ on the tangent bundle of M are said to be *projectively equivalent* if there exists a smooth 1-form ω so

$$\nabla_X Y = \tilde{\nabla}_X Y + \omega(X)Y + \omega(Y)X \quad \text{for all} \quad X, Y;$$

ω is said to provide a projective equivalence from $\tilde{\nabla}$ to ∇. If, in addition, ω is closed, then ∇ and $\tilde{\nabla}$ are said to be *strongly projectively equivalent*. If ∇ is projectively equivalent (resp. strongly projectively equivalent) to a flat connection (i.e., to a connection which has zero Christoffel symbols in some coordinate atlas), then ∇ is said to be *projectively flat* (resp. *strongly projectively flat*). See, for example, the discussion in Opozda [51].

Theorem 12.11

1. *If ω provides a projective equivalence between ∇ and $\tilde{\nabla}$, then $d\omega = 0$ if and only if $\rho_{a,\nabla} = \rho_{a,\tilde{\nabla}}$.*

2. *If two projectively equivalent connections have symmetric Ricci tensors, then they are strongly projectively equivalent.*

3. *If $\mathcal{M}$ is a simply connected surface, then a connection ∇ is strongly projectively flat if and only if ρ and $\nabla\rho$ are totally symmetric.*

4. *If ∇ is a connection, then a curve with $\dot{C} \neq 0$ is an unparameterized geodesic if and only if $\nabla_{\dot{C}}\dot{C} = \alpha(t)\dot{C}$ for some smooth function α.*

5. *Two connections ∇ and $\tilde{\nabla}$ are projectively equivalent if and only if their unparameterized affine geodesics coincide.*

Proof. Let ω provide a projective equivalence between ∇ and $\tilde{\nabla}$. Let $j \neq k$. We use the identity $\Gamma_{ij}{}^k = \tilde{\Gamma}_{ij}{}^k + \delta_i^k \omega_j + \delta_j^k \omega_i$ and Equation (12.1.a) to compute

$$\rho_{jk} - \rho_{kj} = \partial_{x^k} \Gamma_{ij}{}^i - \partial_{x^j} \Gamma_{ik}{}^i, \quad \tilde{\rho}_{jk} - \tilde{\rho}_{kj} = \partial_{x^k} \tilde{\Gamma}_{ij}{}^i - \partial_{x^j} \tilde{\Gamma}_{ik}{}^i,$$

$$\rho_{jk} - \rho_{kj} - \tilde{\rho}_{jk} + \tilde{\rho}_{kj} = \partial_{x^k}(\Gamma_{ij}{}^i - \tilde{\Gamma}_{ij}{}^i) - \partial_{x^j}(\Gamma_{ik}{}^i - \tilde{\Gamma}_{ik}{}^i)$$

$$= (m+1)\{\partial_{x^k} \omega_j - \partial_{x^j} \omega_k\}.$$

It now follows that $\rho_{a,\nabla} = \rho_{a,\tilde{\nabla}}$ if and only if $d\omega = 0$ and Assertion 1 follows (see also the discussion in Steglich [58]). If ∇ and $\tilde{\nabla}$ have symmetric Ricci tensors, then $\rho_{a,\nabla} = \rho_{a,\tilde{\nabla}} = 0$, so Assertion 2 follows from Assertion 1. We refer to Eisenhart [29] (pages 88–96) for the proof of Assertion 3 (see also Theorem 3.3 in Kobayashi and Nomizu [39]).

We prove Assertion 4 as follows. Let $\sigma(s) = C(t(s))$ reparameterize a curve C. Then $\partial_s \sigma^i = (\partial_s t)\dot{C}^i$ and $\partial_s^2 \sigma^i = (\partial_s^2 t)\dot{C}^i + (\partial_s t)^2 \ddot{C}^i$. Thus:

$$\{\nabla_{\partial_s \sigma} \partial_s \sigma\}^k = \partial_s^2 \sigma^k + (\partial_s \sigma^i)(\partial_s \sigma^j)\Gamma_{ij}{}^k = (\partial_s t)^2\{\ddot{C}^k + \Gamma_{ij}{}^k \dot{C}^i \dot{C}^j\} + (\partial_s^2 t)\dot{C}^k. \quad (12.1.g)$$

Suppose first that σ is an unparameterized affine geodesic. Then there exists $t(s)$ so that C is an affine geodesic, i.e., $\ddot{C}^k + \Gamma_{ij}{}^k \dot{C}^i \dot{C}^j = 0$ for all k. We use Equation (12.1.g) to see

$$\nabla_{\partial_s \sigma} \partial_s \sigma = (\partial_s^2 t)\dot{C} = (\partial_s t)^{-1}(\partial_s^2 t)\partial_s \sigma.$$

Set $\alpha(s) := (\partial_s t)^{-1}(\partial_s^2 t)$. Then $\nabla_{\partial_s \sigma} \partial_s \sigma = \alpha(s)\partial_s \sigma$ as desired. Conversely, if we suppose $\nabla_{\partial_s \sigma} \partial_s \sigma = \alpha(s)\partial_s \sigma$, define the function $t = t(s)$ by solving the ordinary differential equation (ODE) $\partial_s^2 t = \alpha(s)\partial_s t$. Then $\ddot{C}^k + \Gamma_{ij}{}^k \dot{C}^i \dot{C}^j = 0$ so $C(t)$ is an affine geodesic. This proves Assertion 4.

Let $\tilde{\nabla}$ be a second affine connection on M. Let $\Theta_{ij}{}^k := \tilde{\Gamma}_{ij}{}^k - \Gamma_{ij}{}^k$; Θ is tensorial and ω provides a projective equivalence from ∇ to $\tilde{\nabla}$ if and only if $\Theta_{ij}{}^k = \delta_i^k \omega_j + \delta_j^k \omega_i$. We compute:

$$\{\tilde{\nabla}_{\dot{C}} \dot{C} - \nabla_{\dot{C}} \dot{C}\}^k = \Theta_{ij}{}^k \dot{C}^i \dot{C}^j. \quad (12.1.h)$$

Suppose that ω provides a projective equivalence from $\tilde{\nabla}$ to ∇. Let C be a ∇ affine geodesic and let $\alpha := 2\omega_i \dot{C}^i$. We then have

$$\{\tilde{\nabla}_{\dot{C}} \dot{C}\}^k = \Theta_{ij}{}^k \dot{C}^i \dot{C}^j = (\delta_i^k \omega_j + \delta_j^k \omega_i)\dot{C}^i \dot{C}^j = \omega_j \dot{C}^k \dot{C}^j + \omega_i \dot{C}^i \dot{C}^k = \alpha \dot{C}^k$$

and C is an unparameterized $\tilde{\nabla}$ affine geodesic.

Conversely, suppose that every affine geodesic for ∇ is an unparameterized affine geodesic for $\tilde{\nabla}$. Let $\{e_1, \ldots, e_m\}$ be a local frame for the tangent bundle. Fix a point P of the manifold in question. Let C be the ∇ affine geodesic with $C(0) = P$ and $\dot{C}(0) = e_1$. By Assertion 4, $\tilde{\nabla}_{\dot{C}} \dot{C}$

is a multiple of $\dot{C}$. Since $\dot{C}^k(P) = 0$ for $k > 1$, we use Equation (12.1.h) to see $\Theta_{11}{}^k = 0$ for $k > 1$. Of course, the use of the index "1" is illustrative only; it holds for any index and any local basis. We use polarization to establish this identity. Consider the shear

$$e_1(\varepsilon) := e_1 + \varepsilon e_2, \quad e_2(\varepsilon) = e_2, \qquad e_k(\varepsilon) = e_k \quad \text{for} \quad k > 2,$$
$$e^1(\varepsilon) = e^1, \qquad\qquad e^2(\varepsilon) = e^2 - \varepsilon e^1, \quad e^k(\varepsilon) = e^k \quad \text{for} \quad k > 2.$$

If $k > 2$, we have

$$0 = \Theta_{11}{}^2(\varepsilon) = \Theta_{11}{}^2 + \varepsilon(2\Theta_{12}{}^2 - \Theta_{11}{}^1) + O(\varepsilon^2),$$
$$0 = \Theta_{11}{}^k(\varepsilon) = \Theta_{11}{}^k + 2\varepsilon\Theta_{12}{}^k + O(\varepsilon^2).$$

Thus $\Theta_{12}{}^k = 0$ and $\Theta_{11}{}^1 = 2\Theta_{12}{}^2$. Permuting the indices yields

$$\Theta_{ij}{}^k = \begin{cases} 0 & \text{if } \{i, j, k\} \text{ are distinct} \\ \tfrac{1}{2}\Theta_{ii}{}^i & \text{if } j = k \text{ and } k \neq i \end{cases} .$$

Let $\omega_1 = \tfrac{1}{2}\Theta_{11}{}^1, \ldots, \omega_m := \tfrac{1}{2}\Theta_{mm}{}^m$. It now follows that $\Theta_{ij}{}^k = \delta_j^k \omega_i + \delta_i^k \omega_j$ and Assertion 5 follows. $\qquad\qquad\square$

12.1.7 THE HESSIAN AND THE AFFINE QUASI-EINSTEIN OPERATOR.

If f belongs to $C^\infty(M)$, then the *Hessian* $\mathcal{H}f = \nabla^2 f = \nabla df$ is the symmetric $(0, 2)$-tensor

$$\mathcal{H}f := \nabla^2 f = (\partial_{x^i}\partial_{x^j} f - \Gamma_{ij}{}^k \partial_{x^k} f)\, dx^i \otimes dx^j .$$

Let $\mathfrak{Q}_{\mu,\nabla} f := \mathcal{H}f - \mu f \rho_s$ be the *affine quasi-Einstein operator*:

$$\mathfrak{Q}_{\mu,\nabla} f = \left\{ \frac{\partial^2 f}{\partial x^i \partial x^j} - \sum_{k=1}^m \Gamma_{ij}{}^k \frac{\partial f}{\partial x^k} - \mu f(\rho_s)_{ij} \right\} dx^i \otimes dx^j . \qquad (12.1.i)$$

The eigenvalue μ is a parameter of the theory. Let

$$E(\mu, \nabla) := \ker\{\mathfrak{Q}_{\mu,\nabla}\} = \{f \in C^2(M) : \mathcal{H}f = \mu f \rho_s\} .$$

Note that if $\rho_s = 0$, then $E(\mu, \nabla) = E(0, \nabla)$ for any μ. Also observe that $E(0, \nabla) = \ker\{\mathcal{H}\}$ is the space of *affine gradient Yamabe solitons*, which we also denote by $\mathcal{Y}$. The equation

$$\mathfrak{Q}_{\mu,\nabla} f = 0$$

is called the *affine quasi-Einstein equation*. We will motivate this equation subsequently when we use the modified Riemannian extension to relate this equation to the quasi-Einstein equation on the tangent bundle in Section 15.1.5 (see Theorem 15.9). We shall study this equation in more detail in Section 12.3. The eigenvalue $\mu = -\frac{1}{m-1}$ is distinguished. If $\mathcal{M} = (M, \nabla)$, we set

$$\mathcal{Q}(\mathcal{M}) := E(-\tfrac{1}{m-1}, \nabla) = \{f \in C^2(M) : \mathcal{H}f = -\tfrac{1}{m-1} f \rho_s\} .$$

This solution space will play an important role in the study of strongly projectively flat geometries.

In Theorem 15.8, we will relate the affine quasi-Einstein equation on an affine surface to the corresponding quasi-Einstein equation for an associated pseudo-Riemannian manifold of neutral signature $(2, 2)$; it is this relationship which motivates our notation for surfaces.

12.1.8 GRADIENT RICCI AND YAMABE SOLITONS.   If (M, g) is a pseudo-Riemannian manifold, and if f is a smooth function on M, one says that (M, g, f) is a *pseudo-Riemannian gradient Yamabe soliton* if there exists a real constant λ so that $\mathcal{H}f = (\tau - \lambda)g$ where τ is the scalar curvature (see Chapter 11 of Book III). One says that (M, g, f) is a *pseudo-Riemannian gradient Ricci soliton* if there exists a real constant so $\mathcal{H}f + \rho = \lambda g$. We refer to Book III for further details concerning these matters.

12.1.9 AFFINE GRADIENT RICCI AND YAMABE SOLITONS.   Many, but not all, notions of Riemannian geometry generalize to the affine category. However, in the affine setting, there is no metric. Consequently, it is not possible to define the scalar curvature, and one proceeds a bit differently. Let $\mathcal{M} = (M, \nabla)$ be an affine manifold. One says that (M, ∇, f) is an *affine gradient Yamabe soliton* if $\mathcal{H}f = 0$, i.e., if f is a solution of the affine quasi-Einstein equation with eigenvalue $\mu = 0$. More generally, one says that (M, ∇, f) is an *affine gradient Ricci soliton* if one has that $\mathcal{H}f + \rho_s = 0$. In local coordinates, this is equivalent to the following:

$$\frac{\partial^2 f}{\partial x^i \partial x^j}(\vec{x}) - \sum_{k=1}^{m} \Gamma_{ij}{}^k(\vec{x})\frac{\partial f}{\partial x^k}(\vec{x}) = -(\rho_s)_{ij}(\vec{x}) \quad \text{for all} \quad i, j. \qquad (12.1.\text{j})$$

Let $\mathcal{Y}(\mathcal{M})$ be the vector space of affine gradient Yamabe solitons for $\mathcal{M}$ and let $\mathfrak{A}(\mathcal{M})$ be the set of affine gradient Ricci solitons. If $\mathfrak{A}(\mathcal{M})$ is non-empty, then $\mathfrak{A}(\mathcal{M})$ is an affine space modeled on $\mathcal{Y}(\mathcal{M})$, i.e., if $f \in \mathfrak{A}(\mathcal{M})$ and if $\xi \in \mathcal{Y}(\mathcal{M})$, then $f + \xi \in \mathfrak{A}(\mathcal{M})$. We say that a 3-tensor field $a_{ijk}dx^i \otimes dx^j \otimes dx^k$ is *totally symmetric* if we have the symmetries $a_{ijk} = a_{jik} = a_{ikj}$. Recall that the action of the curvature on the cotangent bundle is given by $R_{ij}(dx^\ell) = -R_{ijk}{}^\ell dx^k$.

Lemma 12.12   *Let $\mathcal{M} = (M, \nabla)$ be an affine manifold.*

1. *If $f \in \mathfrak{A}(\mathcal{M})$ and if $X \in \mathfrak{K}(\mathcal{M})$, then $X(f) \in \mathcal{Y}(\mathcal{M})$.*
2. *If $h \in \ker\{\mathcal{H}\}$, then $R_{ij}(dh) = 0$ for $1 \leq i < j \leq m$.*
3. *If $f \in \mathfrak{A}(\mathcal{M})$ and if $\nabla \rho_s$ is totally symmetric, then $R_{ij}(df) = 0$ for $1 \leq i < j \leq m$.*

Proof. Let f be an affine gradient Ricci soliton and let X be an affine Killing vector field. We have by naturality that $(\Phi_t^X)^* f$ is again an affine gradient Ricci soliton. Since the difference of two affine gradient Ricci solitons belongs to $\ker\{\mathcal{H}\}$, $(\Phi_t^X)^* f - f \in \ker\{\mathcal{H}\}$. Differentiating this relation with respect to t and setting $t = 0$ yields Assertion 1. Assertion 2 follows from the identity $h_{;ijk} - h_{;ikj} = \{R_{kj}(dh)\}_i$. Assertion 3 follows similarly if we assume $\nabla \rho_s$ to be totally symmetric (see Brozos-Vázquez and García-Río [5]).   □

12.1.10 ELLIPTICITY. We now state a simple result dealing with hypoellipticity and analytic hypoellipticity which follows from more general results in partial differential equations. We refer, for example, to Gilkey [30] for the proof of the first assertion and to Christ [17] or Treves [59] for the proof of the second assertion in the following result.

Lemma 12.13   *Let* $\vec{f} = (f_1, \ldots, f_\ell)$ *be a* C^2 *function which is defined on an open subset* $\mathcal{O}$ *of* $\mathbb{R}^m$. *Suppose that* $\vec{f}$ *satisfies a partial differential equation of the form*

$$\sum_{i=1}^{m} \frac{\partial^2 f_k}{\partial x^i \partial x^i}(\vec{x}) + \sum_{i,j=1}^{m} a_{ijk}(\vec{x}) \frac{\partial f_i}{\partial x^j}(\vec{x}) + \sum_{i=1}^{m} b_{ik}(\vec{x}) f_i(\vec{x}) = c_k(\vec{x}) \quad \text{for} \quad 1 \le k \le \ell.$$

1. *If* $\{a_{ijk}, b_{ik}, c_k\}$ *are smooth, then* $\vec{f}$ *is smooth.*

2. *If* $\{a_{ijk}, b_{ik}, c_k\}$ *are real analytic, then* $\vec{f}$ *is real analytic.*

So far, we have worked with global objects. We let $\mathfrak{K}_P(\mathcal{M})$ be the Lie algebra of *germs* of affine Killing vector fields based at P. The spaces $E_P(\mu, \nabla)$, $\mathcal{Y}_P(\mathcal{M})$, and $\mathfrak{A}_P(\mathcal{M})$ are defined similarly. Let $X \in \mathfrak{K}(\mathcal{M})$. If $F \in \mathfrak{K}(\mathcal{M})$, let $X(F) = [X, F]$. If $f \in E(\mu, \mathcal{M})$, we may apply X to define Xf. We use Lemma 12.13 to establish the following result which we have already used in the proof of Lemma 12.8.

Lemma 12.14   *Let* $\mathcal{M}$ *be a connected real analytic affine manifold.*

1. *Let* $\mathfrak{V} \in \{\mathfrak{K}(\mathcal{M}), \mathfrak{A}(\mathcal{M}), E(\mu, \mathcal{M})\}$.

   (a) *The elements of* $\mathfrak{V}$ *are real analytic.*

   (b) *If* $\mathcal{M}$ *is simply connected and if* $\dim(\mathfrak{V}_P)$ *is constant on* $\mathcal{M}$, *then every element of* $\mathfrak{V}_P$ *extends uniquely to* $\mathfrak{V}$.

   (c) *If* $X \in \mathfrak{V}$ *satisfies* $X(P) = 0$ *and* $\nabla X(P) = 0$, *then* X *vanishes identically.*

2. $\dim(\mathfrak{K}(\mathcal{M})) \le m(m+1)$, $\dim(\mathfrak{A}(\mathcal{M})) \le m+1$, *and* $\dim(E(\preceq, \mathcal{M})) \le m+1$.

3. *Let* $\mathfrak{V} \in \{\mathfrak{K}(\mathcal{M}), E(\mu, \mathcal{M})\}$. *If* $X \in \mathfrak{K}(\mathcal{M})$ *and if* $\xi \in \mathfrak{V}$, *then* $X(\xi) \in \mathfrak{V}$.

Proof. Let $X = a^i \partial_{x^i} \in \mathfrak{K}(\mathcal{M})$. Fix k, set $i = j$ and sum over i in Equation (12.1.d) to see

$$0 = \sum_{i=1}^{m} \left\{ \frac{\partial^2 a^k}{\partial x^i \partial x^i}(\vec{x}) + \sum_{\ell} \left\{ a^\ell(\vec{x}) \frac{\partial \Gamma_{ii}{}^k}{\partial x^\ell}(\vec{x}) - \Gamma_{ii}{}^\ell(\vec{x}) \frac{\partial a^k}{\partial x^\ell}(\vec{x}) + \Gamma_{i\ell}{}^k(\vec{x}) \frac{\partial a^\ell}{\partial x^i}(\vec{x}) + \Gamma_{\ell i}{}^k(\vec{x}) \frac{\partial a^\ell}{\partial x^i}(\vec{x}) \right\} \right\}$$

for all k. We may now use Lemma 12.13 to see X is real analytic. If $f \in E(\mu, \mathcal{M})$, set $i = j$ and sum over i in Equation (12.1.i):

$$0 = \sum_{i=1}^{m} \left\{ \frac{\partial^2 f}{\partial x^i \partial x^i}(\vec{x}) - \mu f(\vec{x})(\rho_s)_{ii}(\vec{x}) - \sum_{k=1}^{m} \Gamma_{ii}{}^k(\vec{x}) \frac{\partial f}{\partial x^k}(\vec{x}) \right\}.$$

Again, we may use Lemma 12.13 to see f is real analytic. If $f \in \mathfrak{A}(\mathcal{M})$, we set $i = j$ and sum over i in Equation (12.1.j) to see

$$\sum_{i=1}^{m} \left\{ \frac{\partial^2 f}{\partial x^i \partial x^i}(\vec{x}) - \sum_{k=1}^{m} \Gamma_{ii}{}^{k}(\vec{x}) \frac{\partial f}{\partial x^k}(\vec{x}) \right\} = - \sum_{i=1}^{m} (\rho_s)_{ii}(\vec{x})$$

and conclude f is real analytic. This proves Assertion 1-a. Assertion 1-b follows by analytic continuation. The defining equations permit us to solve recursively for the second and higher derivatives in terms of the value at P and the values of the derivatives at P. Assertion 1-c now follows as we are in the real analytic category. Assertion 2 follows from Assertion 1-c. Assertion 3 is immediate since the 1-parameter flow defined by an affine Killing vector field preserves the connection ∇ and since the partial differential equation (PDE) in question is homogeneous. □

Remark 12.15   In fact, all the assertions except 1-a continue to hold in the smooth category; see, for example, Brozos-Vázquez et al. [12] for a discussion of $E(\mu, \mathcal{M})$ or Kobayashi and Nomizu [39, Chapter VI] for a discussion of $\mathfrak{K}(\mathcal{M})$.

12.2 SURFACES WITH RECURRENT RICCI TENSOR

In Section 12.2.1 we define recurrence and in Section 12.2.2 we give Wong's classification of recurrent surfaces. In Section 12.2.3 we define the α invariant; this is a local invariant of recurrent surfaces whose Ricci tensor has rank 1 that will play a central role in our discussion of Type $\mathcal{A}$ models subsequently. In Section 12.2.4, we give various examples. Invariant and parallel tensors are defined in Section 12.2.5.

12.2.1 RECURRENCE.   The following result is due to Wong [62].

Theorem 12.16   *Let $\mathcal{M}$ be an affine surface. We say ρ is recurrent if there exists a 1-form ω so $\nabla \rho = \omega \otimes \rho$. Let P be a point of $\mathcal{M}$ such that $\rho(P) \neq 0$. Then any of the two following conditions implies the third:*

1. *There exists a parallel non-trivial vector field near P.*

2. *There exists a parallel non-trivial 1-form near P.*

3. *ρ is recurrent near P.*

In this setting, the recurrence 1-form ω is closed if and only if $\rho_{a,\nabla} = 0$.

Remark 12.17   By Equation (12.1.c), we have $\rho_{11} = R_{211}{}^2$, $\rho_{12} = R_{212}{}^2$, $\rho_{21} = R_{121}{}^1$, and $\rho_{22} = R_{122}{}^1$. Let $\rho_{ij;k} = (\nabla_{\partial_{x^k}} \rho)_{ij}$ and $R_{ijk}{}^{\ell}{}_{;n} = (\nabla_{\partial_{x^n}} R)_{ijk}{}^{\ell}$ be the components of $\nabla \rho$ and ∇R. Lemma 3.5 of Book I shows that given $P \in M$, there exist local coordinates defined near

P so that $\Gamma_{ij}{}^k(P) = 0$. In those coordinates, covariant differentiation and partial differentiation coincide so we may covariantly differentiate these relations to see

$$\rho_{11;i} = R_{211}{}^2{}_{;i}, \quad \rho_{12;i} = R_{212}{}^2{}_{;i}, \quad \rho_{21;i} = R_{121}{}^1{}_{;i}, \quad \rho_{22;i} = R_{122}{}^1{}_{;i}.$$

This shows that ρ is recurrent if and only if R is recurrent.

12.2.2 CLASSIFICATION. A particularly simple example occurs when $\rho_s = 0$ so ρ is a 2-form.

Definition 12.18   Let $\theta = \theta(x^1, x^2)$. Let $\mathcal{R}_0(\theta)$ be the germ of an affine surface where the (possibly) non-zero Christoffel symbols are given by:

$$\mathcal{R}_0(\theta): \quad \Gamma_{11}{}^1 = -\theta^{(1,0)} \quad \text{and} \quad \Gamma_{22}{}^2 = \theta^{(0,1)}.$$

Let $A_0 := \begin{pmatrix} 0 & -1 \\ 1 & 0 \end{pmatrix}$. We compute that

$$\rho = \theta^{(1,1)} A_0, \quad \rho_{;1} = \{\theta^{(1,0)}\theta^{(1,1)} + \theta^{(2,1)}\}A_0, \quad \rho_{;2} = \{\theta^{(1,2)} - \theta^{(0,1)}\theta^{(1,1)}\}A_0.$$

Thus $\rho_s = 0$. Furthermore, if $\theta^{(1,1)} \neq 0$, then the surface is recurrent and the recurrence tensor is given by $\omega = \frac{1}{\theta^{(1,1)}}\{\theta^{(1,0)}\theta^{(1,1)} + \theta^{(2,1)}\}dx^1 + \frac{1}{\theta^{(1,1)}}\{\theta^{(1,2)} - \theta^{(0,1)}\theta^{(1,1)}\}dx^2$.

We have the following result of Derdzinski [25] (see Theorem 6.1) which simplified and extended a previous result of Wong [62] (see Theorem 4.2).

Theorem 12.19   *If P is a point of an affine surface $\mathcal{M}$, then $\rho_s = 0$ near P if and only if $\mathcal{M}$ is locally affine isomorphic near P to the structure $\mathcal{R}_0(\theta)$ for some θ.*

There is a similar classification result if $\rho_s \neq 0$. We first establish some notation. Let $\theta = \theta(x^1, x^2)$. Let $\Delta\theta := \theta^{(2,0)} + \theta^{(0,2)}$. Let $\mathcal{R}_i(\theta)$ be the germ of an affine surface where the (possibly) non-zero Christoffel symbols are given by:

$\mathcal{R}_1(\theta):$ $\Gamma_{11}{}^1 = \Gamma_{12}{}^2 = \theta^{(0,1)}$ and $\Gamma_{11}{}^2 = \theta^{(1,0)} - \theta^{(0,1)}$ for $\theta^{(0,2)} \neq 0$.

$\mathcal{R}_2(\theta):$ $\Gamma_{11}{}^2 = \theta$ for $\theta^{(0,1)} \neq 0$.

$\mathcal{R}_3(\theta):$ $\Gamma_{22}{}^2 = \theta$ for $\theta^{(1,0)} \neq 0$.

$\mathcal{R}_4(\theta):$ $\Gamma_{11}{}^1 = (1+c)\theta^{(1,0)}$, $\Gamma_{22}{}^2 = (1-c)\theta^{(0,1)}$ for $c \notin \{0, \pm 1\}$ and $\theta^{(1,1)} \neq 0$.

$\mathcal{R}_5(\theta):$ $\Gamma_{11}{}^1 = \theta^{(1,0)}$ and $\Gamma_{22}{}^2 = \theta^{(0,1)}$ for $\theta^{(1,1)} \neq 0$.

$\mathcal{R}_6(\theta):$ $\Gamma_{11}{}^1 = \Gamma_{12}{}^2 = -\Gamma_{22}{}^1 = \theta^{(1,0)} + c\theta^{(0,1)}$
      and $\Gamma_{22}{}^2 = \Gamma_{12}{}^1 = -\Gamma_{11}{}^2 = -c\theta^{(1,0)} + \theta^{(0,1)}$ for $c \neq 0$ and $\Delta\theta \neq 0$.

$\mathcal{R}_7(\theta)$: $\Gamma_{11}{}^1 = \Gamma_{12}{}^2 = -\Gamma_{22}{}^1 = \theta^{(1,0)}$ and $\Gamma_{22}{}^2 = \Gamma_{12}{}^1 = -\Gamma_{11}{}^2 = \theta^{(0,1)}$ for $\Delta\theta \neq 0$.

Lemma 12.20 *The structures $\mathcal{R}_i(\cdot)$ given above are recurrent.*

Proof. Set

$$A_1 := \begin{pmatrix} 1 & -1 \\ 1 & 0 \end{pmatrix}, \qquad A_2 := \begin{pmatrix} 1 & 0 \\ 0 & 0 \end{pmatrix}, \qquad A_3 := \begin{pmatrix} 0 & 1 \\ 0 & 0 \end{pmatrix},$$

$$A_4(c) := \begin{pmatrix} 0 & 1-c \\ 1+c & 0 \end{pmatrix}, \qquad A_5 := \begin{pmatrix} 0 & 1 \\ 1 & 0 \end{pmatrix}, \qquad A_6(c) := \begin{pmatrix} 1 & -c \\ c & 1 \end{pmatrix},$$

$$A_7 := \begin{pmatrix} 1 & 0 \\ 0 & 1 \end{pmatrix},$$

$$\xi_{1,1} := 2\theta^{(0,1)}\theta^{(0,2)} - \theta^{(1,2)},$$
$$\xi_{3,2} := \theta\theta^{(1,0)} - \theta^{(1,1)},$$
$$\xi_{4,1} := (c+1)\theta^{(1,0)}\theta^{(1,1)} - \theta^{(2,1)},$$
$$\xi_{4,2} := (1-c)\theta^{(0,1)}\theta^{(1,1)} - \theta^{(1,2)},$$
$$\xi_{5,1} := \theta^{(1,0)}\theta^{(1,1)} - \theta^{(2,1)},$$
$$\xi_{5,2} := \theta^{(0,1)}\theta^{(1,1)} - \theta^{(1,2)},$$
$$\xi_{6,1} := 2c\theta^{(0,1)}\Delta\theta + 2\theta^{(0,2)}\theta^{(1,0)} + 2\theta^{(2,0)}\theta^{(1,0)} - \theta^{(1,2)} - \theta^{(3,0)},$$
$$\xi_{6,2} := -2c\theta^{(0,2)}\theta^{(1,0)} - 2c\theta^{(1,0)}\theta^{(2,0)} - \theta^{(0,3)} + 2\theta^{(0,1)}\Delta\theta - \theta^{(2,1)},$$
$$\xi_{7,1} := -\theta^{(1,2)} + 2\theta^{(1,0)}\Delta\theta - \theta^{(3,0)},$$
$$\xi_{7,2} := -\theta^{(0,3)} + 2\theta^{(0,1)}\Delta\theta - \theta^{(2,1)}.$$

Let $\rho_{;1} := \nabla_{\partial_{x^1}}\rho$ and $\rho_{;2} := \nabla_{\partial_{x^2}}\rho$. We compute:

$$\begin{aligned}
\mathcal{R}_1(\theta): &\quad \rho = -\theta^{(0,2)}\,A_1, &\quad \rho_{;1} = \xi_{1,1}A_1, &\quad \rho_{;2} = -\theta^{(0,3)}\,A_1, \\
\mathcal{R}_2(\theta): &\quad \rho = \theta^{(0,1)}\,A_2, &\quad \rho_{;1} = \theta^{(1,1)}A_2, &\quad \rho_{;2} = \theta^{(0,2)}A_2, \\
\mathcal{R}_3(\theta): &\quad \rho = -\theta^{(1,0)}\,A_3, &\quad \rho_{;1} = -\theta^{(2,0)}A_3, &\quad \rho_{;2} = \xi_{3,2}\,A_3, \\
\mathcal{R}_4(\theta): &\quad \rho = -\theta^{(1,1)}\,A_4, &\quad \rho_{;1} = \xi_{4,1}A_4, &\quad \rho_{;2} = \xi_{4,2}A_4, \\
\mathcal{R}_5(\theta): &\quad \rho = -\theta^{(1,1)}\,A_5, &\quad \rho_{;1} = \xi_{5,1}A_5, &\quad \rho_{;2} = \xi_{5,2}A_5, \\
\mathcal{R}_6(\theta): &\quad \rho = -\Delta\theta\,A_6, &\quad \rho_{;1} = \xi_{6,1}A_6, &\quad \rho_{;2} = \xi_{6,2}A_6, \\
\mathcal{R}_7(\theta): &\quad \rho = -\Delta\theta\,A_7, &\quad \rho_{;1} = \xi_{7,1}A_7, &\quad \rho_{;2} = \xi_{7,2}A_7.
\end{aligned}$$

The result now follows. □

The following result is due to Wong [62] (see Theorem 4.1).

Theorem 12.21 *Let P be a point of an affine surface $\mathcal{M}$ with recurrent Ricci tensor and with $\rho_s(P) \neq 0$. Then $\mathcal{M}$ is locally affine isomorphic near P to one of the structures $\mathcal{R}_i(\cdot)$ given above.*

12.2.3 LOCAL INVARIANTS. In general, it can be difficult to define local invariants in the affine setting since we do not have a metric tensor to raise and lower indices. The following local invariants are not of Weyl type; they do not arise from contracting indices. Other local invariants will be constructed subsequently in Section 15.2 in our discussion of VSI manifolds. Let ρ_s be the symmetric Ricci tensor as defined in Equation (12.1.b).

Lemma 12.22   *Let P be a point of an affine surface M with recurrent Ricci tensor. Assume that* $\mathrm{Rank}(\rho_s) = 1$ *and that* $\nabla\rho_s$ *is totally symmetric. Choose a vector field X so that $\rho_s(X, X) \neq 0$. Let* $\alpha_X := \rho_s(X, X)^{-3}\nabla\rho_s(X, X; X)^2$. *Then α_X is independent of the particular X chosen and setting* $\alpha := \alpha_X$ *defines a smooth local invariant of M.*

Proof. Let $\ker\{\rho_s\} := \{\xi : \rho_s(\xi, \eta) = 0 \text{ for all } \eta\}$. By assumption, $\dim(\ker\{\rho_s\}) = 1$. We work locally and find a smooth section ξ to this 1-dimensional distribution. Choosing local coordinates so $\partial_{x^2} = \xi$, we have

$$(\rho_s)_{11} = \theta_1, \qquad (\rho_s)_{12} = 0, \qquad (\rho_s)_{22} = 0,$$
$$(\rho_s)_{11;1} = \omega_1(\rho_s)_{11}, \quad (\rho_s)_{12;1} = \omega_1(\rho_s)_{12} = 0, \quad (\rho_s)_{22;1} = \omega_1(\rho_s)_{22} = 0,$$
$$(\rho_s)_{11;2} = \omega_2(\rho_s)_{11}, \quad (\rho_s)_{12;2} = \omega_2(\rho_s)_{12} = 0, \quad (\rho_s)_{22;2} = \omega_2(\rho_s)_{22} = 0.$$

Since $\nabla\rho_s$ is totally symmetric, $(\rho_s)_{11;2} = (\rho_s)_{12;1} = 0$. This shows that $\rho_s = \theta_1 dx^1 \otimes dx^1$ and that $\nabla\rho_s = \omega_1\theta_1 dx^1 \otimes dx^1 \otimes dx^1$. Express $X = a_1\partial_{x^1} + a_2\partial_{x^2}$. Then

$$\alpha_X = \frac{\nabla\rho_s(X, X; X)^2}{\rho_s(X, X)^3} = \frac{a_1^6\omega_1^2\theta_1^2}{a_1^6\theta_1^3} = \frac{\omega_1^2}{\theta_1}$$

is independent of X. The desired result now follows.   □

12.2.4 EXAMPLES. The proof of Lemma 12.20 exhibits the Ricci tensors and the covariant derivatives of the structures $\mathcal{R}_i(\theta)$; only $\mathcal{R}_1(\theta)$ and $\mathcal{R}_2(\theta)$ give rise Ricci tensors where $\mathrm{Rank}(\rho_s) = 1$.

Example 12.23   Suppose $M = \mathcal{R}_1(\theta)$. Then $\rho = -\theta^{(0,2)}A_1$ so ∂_{x^2} spans $\ker\{\rho_s\}$. We have $\rho_{;2} = -\theta^{(0,3)}A_1$ so $\nabla\rho_s$ is totally symmetric if and only if $\theta^{(0,3)} = 0$. This means that we can express θ in the form $\theta = a_0(x^1) + a_1(x^1)x^2 + a_2(x^1)(x^2)^2$. We impose this condition and assume a_2 does not vanish identically. We compute

$$\alpha = -\frac{\left(4a_2(x^1)(a_1(x^1) + 2x^2 a_2(x^1)) - 2a_2'(x^1)\right)^2}{8a_2(x^1)^3}.$$

Since α exhibits non-trivial x^2 dependence and since α is an affine invariant, this example is not homogeneous. If a_1 and a_2 are constant, and $a_0''(x^1) = 0$, then ∂_{x^1} is an affine Killing vector field so there are *cohomogeneity 1* examples in this family.

Example 12.24   Suppose $\mathcal{M} = \mathcal{R}_2(\theta)$ so ∂_{x^2} spans $\ker\{\rho_s\}$. We have $\rho = \theta^{(0,1)} A_2$. We have $\rho_{;1} = \theta^{(1,1)} A_2$, and $\rho_{;2} = \theta^{(0,2)} A_2$. Thus $\nabla \rho_s$ is totally symmetric if and only if $\theta^{(0,2)} = 0$. This means that we can express θ in the form $\theta = a_0(x^1) + a_1(x^1) x^2$. We compute

$$\alpha = \frac{a_1'(x^1)^2}{a_1(x^1)^3}.$$

This is constant if and only if $a_1(x^1) = c_0(x^1 - c_1)^{-2}$.

12.2.5 INVARIANT AND PARALLEL TENSORS OF TYPE (1,1).

Let $\mathcal{M}$ be a locally affine homogeneous manifold and let T be a smooth endomorphism of TM, i.e., a tensor of Type (1,1). We expand

$$T = T^i{}_j \partial_{x^i} \otimes dx^j \in C^\infty(TM \otimes T^*M) \quad \text{where} \quad T\partial_{x^j} = \sum_i T^i{}_j \partial_{x^i}.$$

We say that T is an *invariant tensor of Type (1,1)* if $\mathcal{L}_X T = 0$ for all $X \in \mathfrak{K}(\mathcal{M})$. If T is invariant, then $\mathrm{Tr}\{T\}$ is constant. Since Id is invariant, we may assume T is trace-free. We will investigate this subsequently in Section 13.2.8 and in Section 14.2.6 in the context of affine homogeneous surfaces. Let $X = a^k \partial_{x^k}$. We have $\mathcal{L}_X(\partial_{x^i}) = [X, \partial_{x^i}] = -(\partial_{x^i} a^k)\partial_{x^k}$ and dually we obtain $\mathcal{L}_X(dx^j) = \partial_{x^k} a^j dx^k$. Thus

$$
\begin{aligned}
\mathcal{L}_X T &= (a^k \partial_{x^k} T^i{}_j)\partial_{x^i} \otimes x^j - T^i{}_j (\partial_{x^i} a^k)\partial_{x^k} \otimes dx^j + T^i{}_j (\partial_{x^k} a^j)\partial_{x^i} \otimes dx^k \\
&= \{a^k \partial_{x^k} T^i{}_j - T^k{}_j (\partial_{x^k} a^i) + T^i{}_k (\partial_{x^j} a^k)\}\partial_{x^i} \otimes dx^j.
\end{aligned}
$$

Thus $\mathcal{L}_X T = 0$ if and only if we have the equations

$$a^k \partial_{x^k} T^i{}_j - T^k{}_j (\partial_{x^k} a^i) + T^i{}_k (\partial_{x^j} a^k) \quad \text{for all} \quad i, j. \tag{12.2.a}$$

This is studied further in Section 13.2.8 and Section 14.2.6.

We refer to Calviño-Louzao et al. [13] for the proof of the remaining results in this section. We say that a tensor field T of Type $(1, 1)$ on $\mathcal{M}$ is *parallel* if $\nabla T = 0$. Let $\mathcal{P}(\mathcal{M})$ be the set of parallel tensors of Type $(1, 1)$ on $\mathcal{M}$:

$$\mathcal{P}(\mathcal{M}) = \{T^i{}_j : \partial_{x^k} T^i{}_j + \Gamma_{k\ell}{}^i T^\ell{}_j - \Gamma_{kj}{}^\ell T^i{}_\ell = 0, \ \forall \, i, j, k\}.$$

The following is a basic observation.

Lemma 12.25   *If $\mathcal{M} = (M, \nabla)$ is a connected affine surface, then $\mathcal{P}(\mathcal{M})$ is a unital algebra with $\dim(\mathcal{P}(\mathcal{M})) \leq 4$. Let $T \in \mathcal{P}(\mathcal{M})$. The eigenvalues of T are constant on M. If T vanishes at any point of M, then T vanishes identically.*

Let $\mathcal{P}^0(\mathcal{M}) := \{T \in \mathcal{P}(\mathcal{M}) : \mathrm{Tr}\{T\} = 0\}$ be the space of trace-free parallel tensors of Type (1,1), where $\mathrm{Tr}\{T\} := T^i{}_i$ is the trace of T. If $T \in \mathcal{P}(\mathcal{M})$, $\mathrm{Tr}\{T\}$ is constant and expressing $T = \frac{1}{2}\mathrm{Tr}\{T\}\,\mathrm{Id} + (T - \frac{1}{2}\mathrm{Tr}\{T\}\,\mathrm{Id})$ decomposes

$$\mathcal{P}(\mathcal{M}) = \mathrm{Id}\cdot\mathbb{R} \oplus \mathcal{P}^0(\mathcal{M})\,.$$

If $0 \neq T \in \mathcal{P}^0(\mathcal{M})$, then the eigenvalues of T are $\{\pm\lambda\}$ so $\mathrm{Tr}\{T^2\} = 2\lambda^2$. If $2\lambda^2 < 0$ (resp. $2\lambda^2 > 0$), we can rescale T so $T^2 = -\,\mathrm{Id}$ (resp. $T^2 = \mathrm{Id}$) and T defines a *Kähler* (resp. *para-Kähler*) structure on M; the almost complex (resp. almost para-complex) structure being integrable as M is a surface (see Cortés et al. [19] and Newlander and Nirenberg [46]). Finally, if $\lambda = 0$, then T is nilpotent and defines what we will call a *nilpotent Kähler structure*. The symmetric Ricci tensor plays a crucial role.

Theorem 12.26   *Let $\mathcal{M} = (M, \nabla)$ be a simply connected affine surface.*

1. *If* $\dim(\mathcal{P}^0(\mathcal{M})) = 1$, *then exactly one of the following possibilities holds:*

    (a) M admits a Kähler structure and $\mathrm{Rank}(\rho_s) = 2$.

    (b) M admits a para-Kähler structure and $\mathrm{Rank}(\rho_s) = 2$.

    (c) M admits a nilpotent Kähler structure and $\mathrm{Rank}(\rho_s) = 1$.

2. $\dim(\mathcal{P}^0(\mathcal{M})) \neq 2$.

3. $\dim(\mathcal{P}^0(\mathcal{M})) = 3$ *if and only if $\rho_s = 0$. This implies M admits Kähler, para-Kähler, and nilpotent Kähler structures.*

Generically, of course, $\dim(\mathcal{P}^0(\mathcal{M})) = 0$. Furthermore, there exist examples with $\mathrm{Rank}(\rho_s) = 1$ or $\mathrm{Rank}(\rho_s) = 2$ where $\dim(\mathcal{P}^0(\mathcal{M})) = 0$. What is somewhat surprising is that the existence of parallel $(1, 1)$ tensor fields is completely characterized by the geometry of the symmetric part of the Ricci tensor ρ_s.

Theorem 12.27   *Let $\mathcal{M} = (M, \nabla)$ be a simply connected affine surface with $\rho_s \neq 0$.*

1. *M admits a Kähler structure if and only if $\det(\rho_s) > 0$ and ρ_s is recurrent.*

2. *M admits a para-Kähler structure if and only if $\det(\rho_s) < 0$ and ρ_s is recurrent.*

3. *M admits a nilpotent Kähler structure if and only if ρ_s is of rank 1 and recurrent.*

12.3   THE AFFINE QUASI-EINSTEIN EQUATION

Lemma 12.14 provides foundational results concerning $E_P(\mu, \nabla)$ that we will use repeatedly. We first use polarization to establish some results relating to the affine quasi-Einstein equation under strong projective equivalence. In what follows, it will be convenient to work with just one component of the affine quasi-Einstein operator. Suppose that Φ is a symmetric $(0, 2)$-tensor defined on some vector space V and suppose that one could show that $\Phi_{11} = 0$ relative to any basis. It then follows that $\Phi = 0$; this process is called *polarization*.

Section 12.3.1 presents foundational results concerning strong projective equivalence. Section 12.3.2 shows that the solution space Q of the affine quasi-Einstein equation corresponding to the distinguished eigenvalue $\mu = -\frac{1}{m-1}$ transforms conformally under strong projective equivalence. Section 12.3.3 discusses the case when Q has maximal dimension $m + 1$. In Section 12.3.4, the solution space Q is used to give an algorithm for constructing affine geodesics for strongly projectively flat manifolds. The role of the alternating Ricci tensor in the study of Q is treated in Section 12.3.5. Yamabe solitons are examined in Section 12.3.6.

12.3.1 STRONG PROJECTIVE EQUIVALENCE.

Let $\omega = dg$ provide a strong projective equivalence from ∇ to $^g\nabla$, i.e., $^g\nabla_X Y = \nabla_X Y + X(g)Y + Y(g)X$.

Lemma 12.28   *Adopt the notation established above.*

1. $\rho_{s,{}^g\nabla} = \rho_{s,\nabla} - (m-1)\{\mathcal{H}_\nabla g - dg \otimes dg\}$.
2. *If* $\mu = -\frac{1}{m-1}$ *or if* $\mathcal{H}_\nabla g - dg \otimes dg = 0$, *then* $e^g \mathfrak{Q}_{\mu,\nabla} e^{-g} = \mathfrak{Q}_{\mu,{}^g\nabla}$.

Proof. By definition, $^g\Gamma_{ij}{}^k = \Gamma_{ij}{}^k + \delta_i^k \partial_{x^j} g + \delta_j^k \partial_{x^i} g$. Fix a point P of M. Since we are working in the category of torsion-free connections, we can choose a coordinate system so $\Gamma(P) = 0$. We compute at the point P and set $\Gamma_{ij}{}^k(P) = 0$. By Equation (12.1.a),

$$
\begin{aligned}
(\rho_{{}^g\nabla})_{11}(P) &= \{\partial_{x^i}{}^g\Gamma_{11}{}^i - \partial_{x^1}{}^g\Gamma_{i1}{}^i + {}^g\Gamma_{in}{}^i {}^g\Gamma_{11}{}^n - {}^g\Gamma_{1n}{}^i {}^g\Gamma_{i1}{}^n\}(P) \\
&= \{\partial_{x^i}\Gamma_{11}{}^i - \partial_{x^1}\Gamma_{i1}{}^i + (1-m)\partial_{x^1 x^1} g \\
&\qquad + 2(m+1)(\partial_{x^1} g)^2 - (m+3)(\partial_{x^1} g)^2\}(P) \\
&= \{(\rho_\nabla)_{11} - (m-1)(\partial_{x^1 x^1} g - (\partial_{x^1} g)^2)\}(P) \\
&= \{\rho_\nabla - (m-1)(\mathcal{H}_\nabla g - dg \otimes dg)\}_{11}(P).
\end{aligned}
$$

We use polarization to establish Assertion 1. To prove Assertion 2, we examine $\mathfrak{Q}_{\mu,\nabla,11}$ and $\{e^{-g}\mathfrak{Q}_{\mu,{}^g\nabla} e^g\}_{11}$ at P. We compute:

$$
\begin{aligned}
\{e^{-g}\mathcal{H}_{{}^g\nabla,11} e^g f\}(P) &= \{e^{-g}\partial_{x^1 x^1}(fe^g) - {}^g\Gamma_{11}{}^k e^{-g}\partial_{x^k}(fe^g)\}(P) \\
&= \{\partial_{x^1 x^1} f + 2\partial_{x^1} f\, \partial_{x^1} g + f\partial_{x^1 x^1} g + f(\partial_{x^1} g)^2 - 2\partial_{x^1} g(\partial_{x^1} f + f\partial_{x^1} g)\}(P) \\
&= \{\mathcal{H}_{\nabla,11} f + f(\partial_{x^1 x^1} g - (\partial_{x^1} g)^2)\}(P).
\end{aligned}
$$

We use Assertion 1 to obtain:

$$
\begin{aligned}
\{e^{-g}\mathfrak{Q}_{\mu,{}^g\nabla} e^g f &- \mathfrak{Q}_{\mu,\nabla} f\}_{11}(P) \\
&= \{e^{-g}(\mathcal{H}_{{}^g\nabla} e^g f - \mu\rho_{s,{}^g\nabla} e^g f)_{11} - (\mathcal{H}_\nabla f - \mu\rho_{s,\nabla} f)_{11}\}(P) \\
&= \{f(1 + (m-1)\mu)(\partial_{x^1 x^1} g - (\partial_{x^1} g)^2)\}(P).
\end{aligned}
$$

We use polarization to establish the following identity from which Assertion 2 follows:

$$
e^{-g}\mathfrak{Q}_{\mu,{}^g\nabla} e^g f - \mathfrak{Q}_{\mu,\nabla} f = (1 + (m-1)\mu) f(\mathcal{H}_\nabla g - dg \otimes dg). \qquad \square
$$

12.3.2 CONFORMAL EQUIVALENCE. Recall that

$$\mathcal{Q}(\mathcal{M}) := \{ f \in C^\infty(M) : \mathcal{H}f + \tfrac{1}{m-1} f \rho_s = 0 \} \, .$$

Theorem 12.29   *If $g \in C^\infty(M)$, let $^g\mathcal{M} = (M, {}^g\nabla)$.*

1. $\mathcal{Q}(^g\mathcal{M}) = e^g \mathcal{Q}(\mathcal{M})$.

2. *The following assertions are equivalent:*

    (a) $\rho_{s,{}^g\nabla} = \rho_{s,\nabla}$.   (b) $\mathcal{H}g - dg \otimes dg = 0$.   (c) $e^{-g} \in E(0, \nabla)$.

3. *If any of the assertions in (2) hold, then $E(\mu, {}^g\nabla) = e^g E(\mu, \nabla)$ for any μ.*

Proof. Assertion 1 is immediate from the intertwining relation of Lemma 12.28. The equiva-lence of Assertion 2-a and Assertion 2-b again follows from Lemma 12.28. The equivalence of Assertion 2-b and Assertion 2-c follows by noting

$$\mathfrak{Q}_{0,\nabla}(e^{-g}) = \mathcal{H}(e^{-g}) = -e^{-g}\{\mathcal{H}(g) - dg \otimes dg\} \, .$$

Assertion 3 now follows from Assertion 2-b and Lemma 12.28.   □

Remark 12.30   Theorem 12.11 and Theorem 12.29 deal with strong projective equivalence. If $\mathcal{M}$ and $\tilde{\mathcal{M}}$ are strongly projectively equivalent, then

$$\rho_{a,\mathcal{M}} = \rho_{a,\tilde{\mathcal{M}}} \quad \text{and} \quad \dim(\mathcal{Q}(\mathcal{M})) = \dim(\mathcal{Q}(\tilde{\mathcal{M}})) \, .$$

However, if $\mathcal{M}$ and $\tilde{\mathcal{M}}$ are only projectively equivalent, then these two identities can fail. Let $\mathcal{M}$ be the usual flat structure on $\mathbb{R}^2$. Let $\omega = x^2 dx^1$ define a projective equivalence from $\mathcal{M}$ to $\tilde{\mathcal{M}}$. Let $P = 0$. A direct computation shows that

$$\dim(\mathcal{Q}_P(\mathcal{M})) = 3, \quad \dim(\mathcal{Q}_P(\tilde{\mathcal{M}})) = 0,$$

$$\rho_\mathcal{M} = \begin{pmatrix} 0 & 0 \\ 0 & 0 \end{pmatrix}, \quad \rho_{\tilde{\mathcal{M}}} = \begin{pmatrix} (x^2)^2 & 1 \\ -2 & 0 \end{pmatrix} \, .$$

We can draw the following conclusion from Theorem 12.29.

Lemma 12.31   *Let $\mathcal{M}$ be an affine manifold of dimension m.*

1. *If $\mu \neq 0$, then $\rho_{s,\mathcal{M}} = 0$ if and only if $\mathbb{1} \in E(\mu, \mathcal{M})$.*

2. *$\mathcal{M}$ is strongly projectively equivalent to an affine surface $\mathcal{M}_1$ with $\rho_{s,\mathcal{M}_1} = 0$ if and only if there exists $f \in \mathcal{Q}_P(\mathcal{M})$ with $f(P) \neq 0$.*

Proof. We have $f \in E_P(\mu, \mathcal{M})$ if and only if $\mathcal{H}f = \mu f \rho_s$. Since $\mathcal{H}\mathbb{1} = 0$, $\mathbb{1} \in E_P(\mu, \mathcal{M})$ if and only if $\mu \rho_s = 0$. Assertion 1 follows since $\mu \neq 0$. Suppose dg provides a local strong projective equivalence from $\mathcal{M}$ to $\mathcal{M}_1$ with $\rho_{s,\mathcal{M}_1} = 0$. By Theorem 12.29, $\mathcal{Q}(\mathcal{M}) = e^{-g}\mathcal{Q}(\mathcal{M}_1)$. By Assertion 1, $\mathbb{1} \in \mathcal{Q}(\mathcal{M}_1)$. Thus $f := e^{-g} \in \mathcal{Q}(\mathcal{M})$ and we obtain as desired that $f(P) \neq 0$.

Suppose there exists $f \in \mathcal{Q}_P(\mathcal{M})$ so $f(P) \neq 0$. By replacing f by $-f$ if necessary, we may assume $f(P) > 0$ and set $g = \log(f)$. Let $\mathcal{M}_1 = e^{-g}\mathcal{M}$; $\mathbb{1} = e^{-g}e^g \in \mathcal{Q}(\mathcal{M}_1)$. Assertion 1 then yields $\rho_{s,\mathcal{M}_1} = 0$. This completes the proof of Assertion 2. □

12.3.3 $\mathrm{DIM}(\mathcal{Q}(\mathcal{M})) = \mathrm{DIM}(\mathcal{M}) + 1$.

The solution space $\mathcal{Q}$ of the affine quasi-Einstein equation corresponding to the eigenvalue $\mu = -\frac{1}{m+1}$ provides a complete system of invariants if the manifold in question is strongly projectively flat. We establish the results in this section that will play a central role in our discussion of Type $\mathcal{A}$ geometries in Chapter 13. We work locally for the most part. We begin with a useful technical result.

Lemma 12.32   *Let $\mathcal{M}$ be an affine manifold of dimension m with $\dim(E_P(\mu, \nabla)) = m + 1$ for some μ.*

1. *There is a basis $\mathcal{B} = \{\phi_0, \phi_1, \dots, \phi_m\}$ for $E_P(\mu, \nabla)$ so*

$$\phi_0(P) = 1, \ d\phi_0(P) = 0, \ and \ for \ 1 \leq i \leq m, \ \phi_i(P) = 0, \ d\phi_i(P) = dx^i. \quad (12.3.a)$$

   Let $g = \log(\phi_0)$ and let $x^i := e^{-g}\phi_i$ for $i \geq 1$. Then $(x^1, \dots, x^m)$ is a system of local coordinates on $\mathcal{M}$ centered at P such that $E_P(\mu, \nabla) = e^g \operatorname{span}\{x^1, \dots, x^m\}$.

2. *$\mathcal{M}$ is locally strongly projectively flat. If $\mu \neq -\frac{1}{m+1}$, then $\rho_s = 0$.*

Proof. Suppose $\dim(E_P(\mu, \nabla)) = m + 1$. Let $\Theta_P(\phi) := (\phi, \partial_{x^1}\phi, \dots, \partial_{x^m}\phi)(P) \in \mathbb{R}^{m+1}$. If $\Theta_P(\phi) = 0$, then $\phi(P) = 0$ and $\nabla\phi(P) = 0$ so $\phi \equiv 0$ by Lemma 12.14. Consequently, Θ_P is an injective map from $E_P(\mu, \nabla)$ to $\mathbb{R}^{m+1}$. Thus, for dimensional reasons, ϕ_P is bijective and we can choose a basis satisfying the normalizations of Equation (12.3.a). The remaining assertions of Assertion 1 now follow. We compute:

$$0 = e^{-g}\left\{\mathfrak{Q}_{\mu,\nabla}(x^k e^g) - x^k \mathfrak{Q}_{\mu,\nabla}(e^g)\right\}_{ij} = e^{-g}\{\mathcal{H}(x^k e^g) - x^k \mathcal{H}(e^g)\}_{ij}$$
$$= \delta^k_j \, \partial_{x^i} g + \delta^k_i \, \partial_{x^j} g - \Gamma_{ij}{}^k.$$

Let $\tilde{\Gamma}_{ij}{}^k \equiv 0$ define $\tilde{\nabla}$. We have $\Gamma_{ij}{}^k = \tilde{\Gamma}_{ij}{}^k + \delta^k_j \, \partial_{x^i} g + \delta^k_i \, \partial_{x^j} g$ so dg is a strong projective equivalence from $\tilde{\nabla}$ to ∇. Consequently, ∇ is strongly projectively flat. By Theorem 12.29, $\mathcal{Q}_P(\mathcal{M}) = e^g \operatorname{span}\{\mathbb{1}, x^1, \dots, x^m\}$ so

$$0 = \left\{\mathfrak{Q}_{\nu,\nabla} - \mathfrak{Q}_{-\frac{1}{m-1},\nabla}\right\}e^g = \left(\nu + \frac{1}{m-1}\right)e^g \rho_s.$$

Thus if $\nu \neq -\frac{1}{m-1}$, $\rho_s = 0$ and Assertion 2 follows. □

We use essentially the same argument to prove the following result; it will play an important role in our analysis of Type $\mathcal{A}$ models in Section 13.3.

Lemma 12.33   *If $\mathcal{M}$ is a strongly projectively flat surface, then*

$$\mathcal{Q}(\mathcal{M}) \neq e^{g(x^1,x^2)} \operatorname{span}\{f_1(x^1), f_2(x^1), f_3(x^1)\}.$$

Proof. Suppose to the contrary that $\mathcal{Q}(\mathcal{M}) = e^{g(x^1,x^2)} \operatorname{span}\{f_1(x^1), f_2(x^1), f_3(x^1)\}$. We argue for a contradiction. Let $\tilde{\nabla} := {}^{-g}\nabla$; by Theorem 12.29,

$$\mathcal{Q}(M, \tilde{\nabla}) = \operatorname{span}\{f_1(x^1), f_2(x^1), f_3(x^1)\}.$$

Define $f = c_1 f_1 + c_2 f_2 + c_3 f_3$. Since $f = f(x^1)$, we may choose $(c_1, c_2, c_3) \neq (0, 0, 0)$ so that $f(P) = 0$ and $df(P) = 0$; thus $f \equiv 0$ and the functions $\{f_1, f_2, f_3\}$ are not linearly independent which is false.   □

The following result is due to Gilkey and Valle-Regueiro [35].

Theorem 12.34   *Let P be a point of an affine manifold $\mathcal{M}$ of dimension m.*

1. *$\mathcal{M}$ is locally strongly projectively flat if and only if $\dim(\mathcal{Q}_P(\mathcal{M})) = m + 1$.*
2. *If $\dim(\mathcal{Q}_P(\mathcal{M})) = m + 1$ and if $\mathbb{1} \in \mathcal{Q}_P(\mathcal{M})$, then $\mathcal{M}$ is flat near P.*
3. *Let M_i be affine manifolds of dimension m with $\dim(\mathcal{Q}(M_i)) = m + 1$. Let Φ be a diffeomorphism from M_1 to M_2. If $\mathcal{Q}(M_1) = \Phi^* \mathcal{Q}(M_2)$, then $M_1 = \Phi^* M_2$.*

Proof. By Lemma 12.14, $\dim(\mathcal{Q}_P(\mathcal{M})) \leq m + 1$ for any affine manifold of dimension m. By Lemma 12.32, if equality holds, then $\mathcal{M}$ is locally strongly projectively flat. Suppose dg provides a local strong projective equivalence between ∇ and a flat connection ∇_f. We can choose local coordinates $(x^1, \dots, x^m)$ near P so that the Christoffel symbols of ∇_f vanish. It is then immediate that $\{\mathbb{1}, x^1, \dots, x^m\} \subset \mathcal{Q}_P(\mathcal{M}_f)$ so $\dim(\mathcal{Q}_P(\mathcal{M}_f)) = m + 1$. By Theorem 12.29, $\mathcal{Q}_P(\mathcal{M}) = e^g \mathcal{Q}_P(\mathcal{M}_f)$ so $\dim(\mathcal{Q}_P(\mathcal{M})) = m + 1$. This proves Assertion 1.

Suppose $\mathbb{1} \in \mathcal{Q}_P(\mathcal{M})$. Let $\mathcal{B}$ be the basis of Lemma 12.32. Since $\{\phi_0 - \mathbb{1}\}(P) = 1$ and $d\{\phi_0 - \mathbb{1}\}(P) = 0$, $\phi_0 \equiv \mathbb{1}$ by Lemma 12.14. Thus $g = 0$ and $\mathcal{Q}_P(\mathcal{M}) = \operatorname{span}\{\mathbb{1}, x^1, \dots, x^m\}$ in the new system of coordinates. Since $\mathcal{H}\mathbb{1} = 0$ and $\mathbb{1} \in \mathcal{Q}_P(\mathcal{M})$, $\rho_s = 0$. As $x^i \in \mathcal{Q}_P(\mathcal{M})$ and $\rho_s = 0$, $0 = \mathcal{H}x^i = \{\partial_j \partial_k x^i - \Gamma_{jk}{}^\ell \partial_{x^\ell} x^i\} dx^j \otimes dx^k = -\Gamma_{jk}{}^i dx^j \otimes dx^k$. Consequently, $\Gamma = 0$ so $\mathcal{M}$ is locally flat which establishes Assertion 2.

By replacing $\mathcal{M}_2$ by $\Phi^* \mathcal{M}_2$, it suffices to prove Assertion 3 under the additional hypothesis that $M_1 = M_2$ and $\mathcal{Q}(M_1) = \mathcal{Q}(M_2)$. We adopt the notation of Lemma 12.32. Let $\mathcal{Q}(M_1) = \mathcal{Q}(M_2) = e^g \operatorname{span}\{\mathbb{1}, x^1, \dots, x^m\}$. Let $\tilde{\nabla}_1 := {}^{-g}\nabla_1$ and $\tilde{\nabla}_2 := {}^{-g}\nabla_2$. By Theorem 12.29, $\mathcal{Q}(\tilde{M}_1) = \mathcal{Q}(\tilde{M}_2) = \operatorname{span}\{\mathbb{1}, x^1, \dots, x^m\}$. The argument given to prove Assertion 2 then shows $\tilde{M}_1 = \tilde{M}_2$ is defined by the flat connection where all the Christoffel symbols vanish in the new coordinate system. Since $M_1 = {}^g \tilde{M}_2$ and $M_2 = {}^g \tilde{M}_2$, we obtain $M_1 = M_2$ and Assertion 3 follows.   □

By Theorem 12.34, $\mathcal{M}$ is strongly projectively flat if and only if $\dim(\mathcal{Q}(\mathcal{M})) = m + 1$. In the category of strongly projective flat affine manifolds of dimension $m + 1$, Theorem 12.34 shows that the transformation $\mathcal{M} \rightsquigarrow \mathcal{Q}(\mathcal{M})$ is a natural embedding of contravariant functors from the category of strongly projective flat affine manifolds of dimension $m + 1$ (where the morphisms are affine diffeomorphisms) into the category of $(m + 1)$-dimensional vector spaces of functions on smooth manifolds of dimension $m + 1$ (where the morphisms are given by pullbacks induced by diffeomorphisms). This latter category is much more tractable.

12.3.4 GEODESICS IN STRONGLY PROJECTIVELY FLAT MANIFOLDS. Let $\mathcal{M}$
be an affine manifold of dimension m. Choose a system of local coordinates centered at a point P of $\mathcal{M}$. Let $\sigma_{\vec{a}}(t)$ be the affine geodesic on $\mathcal{M}$ with $\sigma_{\vec{a}}(0) = P$ and $\dot{\sigma}_{\vec{a}}(0) = \vec{a}$ in $\mathbb{R}^m$. In general, the ordinary differential equation (ODE) defining $\sigma_{\vec{a}}$ is a quadratic system of ODEs in m functions of t. In the strongly projectively flat category, the following result reduces the geodesic equation to an ODE in a single function of t. This will inform our discussion of affine geodesics in Chapter 13 subsequently.

Theorem 12.35   *Let $\mathcal{M}$ be an affine manifold of dimension m with $\dim(\mathcal{Q}_P(\mathcal{M})) = m + 1$. Adopt the notation of Lemma 12.32 and express $\mathcal{Q}_P(\mathcal{M}) = \mathrm{span}\{\phi_0, \ldots, \phi_m\}$ where the ϕ_i satisfy Equation (12.3.a). Let $\Psi = (\phi_1/\phi_0, \ldots, \phi_m/\phi_0)$ define a local diffeomorphism from $(\mathcal{M}, P)$ to $(\mathbb{R}^m, 0)$. There exists the germ of a smooth function $\psi_{\vec{a}}(t)$ with $\psi_{a,b}(0) = 0$ and $\psi'_{a,b}(0) = 1$ so that $\sigma_{a,b}(t) = \Phi^{-1}(\psi_{\vec{a}}(t)\vec{a})$.*

Proof. Let $\mathcal{M}_0$ be the flat affine structure on $\mathbb{R}^m$. The affine geodesics in $\mathcal{M}_0$ are straight lines; they take the form $\sigma(t) := t\vec{a}$. By Theorem 12.11, two projectively equivalent affine manifolds have the same unparameterized affine geodesics. The desired result now follows since e^g provides a strong projective equivalence between $\Psi^*\mathcal{M}_0$ and $\mathcal{M}$. $\qquad\square$

We will be working in the real analytic category (see Theorem 12.52) in our treatment of locally affine homogeneous surface geometries. Thus we can apply analytic continuation to pass from germs of affine geodesics to the affine geodesics themselves. In Lemma 13.6, we will show any Type $\mathcal{A}$ model is strongly projectively flat and thus this analysis pertains by Theorem 12.34. This ansatz will inform our discussion of geodesic completeness in the Type $\mathcal{A}$ setting.

12.3.5 THE ALTERNATING RICCI TENSOR. We use Theorem 12.19 to establish the following result.

Lemma 12.36   *Assume ρ_s vanishes identically near a point P of an affine surface $\mathcal{M}$.*

1. *If ρ_a does not vanish identically near P, then $E_P(\mu, \mathcal{M}) = \mathrm{span}\{\mathbb{1}\}$.*
2. *If ρ_a vanishes identically near P, then $\dim(E_P(\mu, \mathcal{M})) = 3$.*

Proof. Since ρ_s vanishes identically near P, we may use Lemma 12.31 to see that $\parallel$ belongs to $E(\mu, \mathcal{M})$. Suppose $\rho_a(P) \neq 0$. Use Theorem 12.19 to choose local coordinates near P so that $\Gamma_{11}{}^1 = -\partial_{x^1}\phi$ and $\Gamma_{22}{}^2 = \partial_{x^2}\phi$. Since $\rho = -\partial_{x^1}\partial_{x^2}\phi \, dx^1 \wedge dx^2$, $\partial_{x^1}\partial_{x^2}\phi \neq 0$. We have $E_P(\mu, \mathcal{M}) = \ker_P\{\mathcal{H}\}$. If $\mathcal{H}f = 0$, then $0 = \mathcal{H}_{12}f = \partial_{x^1}\partial_{x^2}f$ and, consequently, $f(x^1, x^2) = a_1(x^1) + a_2(x^2)$. Setting $\mathcal{H}f = 0$ then yields the relations

$$
\begin{aligned}
0 &= \mathcal{H}_{11}f = a_1''(x^1) + a_1'(x^1)\partial_{x^1}\phi(x^1, x^2), \\
0 &= \mathcal{H}_{22}f = a_2''(x^2) - a_2'(x^2)\partial_{x^2}\phi(x^1, x^2).
\end{aligned}
$$

Differentiating $\mathcal{H}_{11}f$ with respect to x^2 and $\mathcal{H}_{22}f$ with respect to x^1 yields

$$
a_1'(x^1)\partial_{x^1}\partial_{x^2}\phi(x^1, x^2) = 0 \quad \text{and} \quad a_2'(x^2)\partial_{x^1}\partial_{x^2}\phi(x^1, x^2) = 0.
$$

Since $\rho_a(P) \neq 0$, we have $\partial_{x^1}\partial_{x^2}\phi(x^1, x^2)(P) \neq 0$. Consequently, $a_1' = 0$ and $a_2' = 0$. This shows that f is a multiple of $\parallel$ and $E_P(\mu, \mathcal{M}) = \mathrm{span}\{\parallel\}$.

Suppose next that $\rho_a(P) = 0$ but there is a sequence of points Q_n which converge to P so that $\rho_a(Q_n) \neq 0$. We then have that $E_{Q_n}(\mu, \mathcal{M}) = \mathrm{span}\{\parallel\}$. Suppose that

$$
\dim(E_P(\mu, \mathcal{M})) \neq 1.
$$

Since $\parallel \in E_P(\mu, \mathcal{M})$, we may choose $0 \neq f \in E_P(\mu, \mathcal{M})$ so that $f(P) = 0$; thus by Lemma 12.14, $df(P) \neq 0$. As $f \in E_{Q_n}(\mu, \mathcal{M})$ for n large satisfies $df(Q_n) \neq 0$, this shows $\dim(E_{Q_n}(\mu, \mathcal{M})) \geq 2$, which is false. Thus $\dim(E_P(\mu, \mathcal{M})) = 1$ and Assertion 1 follows.

If $\rho_s = 0$ and $\rho_a = 0$ near P, then $\mathcal{M}$ is flat near P. Consequently, by Theorem 12.34, $\dim(\mathcal{Q}_P(\mathcal{M})) = 3$. Since $\rho_s = 0$, $\mathcal{Q}_P(\mathcal{M}) = \ker_P\{\mathcal{H}\} = E_P(\mu, \mathcal{M})$ for any μ. This proves Assertion 2. $\qquad\square$

We now establish a result of Brozos-Vázquez et al. [12].

Theorem 12.37 *If $\mathcal{M}$ is an affine surface, then $\dim(\mathcal{Q}_P(\mathcal{M})) \neq 2$.*

Proof. We suppose to the contrary that $\dim(\mathcal{Q}_P(\mathcal{M})) = 2$ and argue for a contradiction. Suppose first that $f(P) \neq 0$ for some $f \in \mathcal{Q}_P(\mathcal{M})$. Then $\mathcal{M}$ is locally strongly projectively equivalent to a surface $\mathcal{M}_1$ with $\rho_{s,\mathcal{M}_1} = 0$ by Lemma 12.31. By Lemma 12.36, $\dim(\mathcal{Q}_P(\mathcal{M}_1)) \neq 2$. By Theorem 12.29, $\dim(\mathcal{Q}_P(\mathcal{M}_1)) = \dim(\mathcal{Q}_P(\mathcal{M})) = 2$. This provides the desired contradiction.

Next we suppose $f(P) = 0$ for all $f \in \mathcal{Q}_P(\mathcal{M})$. By Lemma 12.14, $df(P) \neq 0$. Since $\dim(\mathcal{Q}_P(\mathcal{M})) = 2$, we can choose $\{\phi_1, \phi_2\} \subset \mathcal{Q}_P(\mathcal{M})$ so that $d\phi_i(P) = dx^i$. Thus we can change coordinates to ensure $\phi_i = x^i$. If $\tilde{P} \neq P$, then $\dim(\mathcal{Q}_{\tilde{P}}(\mathcal{M})) \geq 2$. Since there is an element of $\mathcal{Q}_{\tilde{P}}(\mathcal{M})$ which is non-zero, $\dim(\mathcal{Q}_{\tilde{P}}(\mathcal{M})) \neq 2$. Consequently, $\dim(\mathcal{Q}_{\tilde{P}}(\mathcal{M})) = 3$. Theorem 12.34 then shows that $\mathcal{M}$ is strongly projectively flat near $\tilde{P}$. By Theorem 12.11, we conclude ρ and $\nabla\rho$ are totally symmetric at $\tilde{P}$. Continuity then implies ρ and $\nabla\rho$ are totally symmetric at P as well. Hence, by Theorem 12.11, $\mathcal{M}$ is strongly projectively flat near P. This implies $\dim(\mathcal{Q}_P(\mathcal{M})) = 3$, which is a contradiction. $\qquad\square$

Remark 12.38 Theorem 12.37 fails if we replace $\dim(\mathcal{Q}_P(\mathcal{M}))$ by $\dim(\mathcal{Q}(\mathcal{M}))$ and if $\mathcal{M}$ is not simply connected. Let $\mathcal{M} = \mathbb{R}^2$ with the standard flat structure, let $\mathcal{M}_2$ be the cylinder with the standard flat structure where we identify

$$(x^1, x^2) \text{ with } (x^1 + 2n_1\pi, x^2)$$

for $n_1 \in \mathbb{Z}$, and let $\mathcal{M}_1$ be the torus with the standard flat structure where we identify (x^1, x^2) with $(x^1 + 2n_1\pi, x^2 + 2n_2\pi)$ for $(n_1, n_2) \in \mathbb{Z}^2$. The holonomy plays a crucial role as

$$\mathcal{Q}(\mathcal{M}) = \text{span}\{\text{1}, x^1, x^2\}, \quad \mathcal{Q}(\mathcal{M}_1) = \text{span}\{\text{1}, x^2\}, \quad \mathcal{Q}(\mathcal{M}_2) = \text{span}\{\text{1}\}.$$

12.3.6 YAMABE SOLITONS.

The space of Yamabe solitons plays an important role in the study of flat connections. Let Γ_0 be the flat connection all of whose Christoffel symbols vanish identically. The following observation is an immediate consequence of Theorem 12.34 since $\mathcal{Q}(\mathcal{M}) = \mathcal{Y}(\mathcal{M})$ in the flat setting.

Lemma 12.39 Let Γ define a flat connection on $\mathbb{R}^2$. Then $\dim(\mathcal{Y}(\Gamma)) = 3$. Let $F = (f^1, f^2)$ be the germ of a diffeomorphism. We have

1. $F^*\Gamma_0 = \Gamma$ if and only if $\{1, f^1, f^2\}$ is a basis for $\mathcal{Y}(\Gamma)$.
2. $F^*\Gamma = \Gamma$ if and only if $F^*\{\mathcal{Y}(\Gamma)\} = \mathcal{Y}(\Gamma)$.

The following will be a useful observation.

Lemma 12.40 Let $\mathcal{M}$ be an affine manifold of dimension m with $\dim(\mathcal{Y}(\mathcal{M})) \geq m + 1$. Then $\mathcal{M}$ is flat and $\dim(\mathcal{Y}(\mathcal{M})) = m + 1$.

Proof. We apply Lemma 12.14 to see $\dim(\mathcal{Y}(\mathcal{M})) \leq m + 1$, and hence equality holds. We have $\text{1} \in \mathcal{Y}(\mathcal{M})$. Consequently, we may apply Lemma 12.32 with $\mu = 0$ to choose local coordinates so $\mathcal{Y}(\mathcal{M}) = \text{span}\{1, x^1, \ldots, x^m\}$. We have $0 = \mathcal{H}(x^i) = \Gamma_{jk}{}^i dx^j \otimes dx^j$. Consequently, all the Christoffel symbols vanish in this coordinate system. This implies the connection in question is flat. $\square$

12.4 THE CLASSIFICATION OF LOCALLY HOMOGENEOUS AFFINE SURFACES WITH TORSION

We permit the surface in question to have torsion throughout this section as the analysis is no more difficult in this setting. The approach we will take will be quite different in flavor from the proof of the classification result given by Opozda [53] in the torsion-free setting; her work rested upon a detailed analysis of the curvature tensor. A subsequent proof was given by Kowalski, Opozda, and Vlášek [44] which was group theoretic in nature. The classification was completed by Arias-Marco and Kowalski [2] by permitting the surface in question to have torsion. Their work involved a careful examination of the classification of the Lie algebras of vector fields due to Olver [49]. By contrast, the proof we shall give mixes Lie theoretic methods, specifically an analysis of root systems, and a detailed examination of the eight affine Killing Equations (12.1.d). We also refer to related work of Guillot and Sánchez-Godinez [37], Kowalski, Opozda, and Vlášek [42], and Opozda [54].

    T. Arias-Marco   O. Kowalski   P. Olver[1]   B. Opozda

There is always a question of the local versus the global geometry of an object in Differential Geometry. Let $\mathcal{M}$ be a locally affine homogeneous surface. By Lemma 12.14, for the objects under consideration, questions of passing from the local to the global involve the holonomy action of the fundamental group; there is no obstruction if $\mathcal{M}$ is assumed simply connected. We shall not belabor the point and ignore the question of passing from local to global henceforth for the most part.

Choose a reference point $P \in M$; any other point will do as well since we have assumed that $\mathcal{M}$ is locally affine homogeneous and connected. Thus we can assume without loss of generality that M is an arbitrarily small neighborhood of P. Consequently, for example, if f is a smooth function with $f(P) \neq 0$, we may assume f never vanishes on M. Similarly, if X_1 and X_2 are two vector fields which are linearly independent at P, we may assume that X_1 and X_2 are linearly independent on all of M.

Here is a brief outline to this section. In Section 12.4.1, we present the four families of locally homogeneous pseudo-Riemannian surfaces. In Section 12.4.2, we discuss the two simply connected 2-dimensional Lie groups. These are the additive group $\mathbb{R}^2$, which is Abelian, and the $ax + b$ group, which is non-Abelian. Denote the associated Lie algebras of these groups by $\mathfrak{K}_A$ and $\mathfrak{K}_B$, respectively. Let $\mathfrak{so}(3)$ be the Lie algebra of the 3-dimensional special orthogonal group SO(3). A Lie subalgebra of the Lie algebra of affine Killing vector fields is said to be *effective* if it spans the tangent space of M at some point. In Section 12.4.3, we discuss affine surface geometries with torsion where $\mathfrak{K}(\mathcal{M})$ contains an effective Lie subalgebra isomorphic to $\mathfrak{K}_A$, $\mathfrak{K}_B$, or $\mathfrak{so}(3)$. In Section 12.4.4, we discuss the exceptional Lie algebra $A_{3,7}^a$.

In Section 12.4.5, we use the adjoint action of $\mathfrak{K}(\mathcal{M})$ on itself to show that $\mathfrak{K}(\mathcal{M})$ must contain an effective Lie subalgebra isomorphic to $\mathfrak{K}_A$, to $\mathfrak{K}_B$ or to $\mathfrak{so}(3)$. These results are then

---

[1]Photo credit University of Minnesota

used in Section 12.4.6 to complete the classification of locally homogeneous surfaces with torsion in Theorem 12.49. Metrizability is treated briefly in Section 12.4.7.

12.4.1 LOCALLY HOMOGENEOUS PSEUDO-RIEMANNIAN SURFACES.

A pseudo-Riemannian manifold is said to be *locally homogeneous* if given any two points of the manifold, there is a local isometry taking one point to the other. An affine manifold is said to be *locally affine homogeneous* if given any two points of the manifold there is a local affine map taking one point to the other. We delete the word "local" if the maps in question are global diffeomorphisms.

A pseudo-Riemannian surface is locally homogeneous if and only if it has constant scalar curvature. Up to a non-necessarily positive rescaling and isometry, there are four simply connected geodesically complete homogeneous pseudo-Riemannian geometries:

1. Let $\mathbb{E}^2$ be $\mathbb{R}^2$ with the flat metric $ds^2 = (dx^1)^2 + (dx^2)^2$.

2. Let $\mathfrak{H}$ be the unit sphere in $\mathbb{R}^3$ with the round metric.

3. Let $\mathbb{H}$ be the upper half-plane $\mathbb{R}^+ \times \mathbb{R}$ with the metric $ds^2 = (x^1)^{-2}((dx^1)^2 + (dx^2)^2)$.

4. Let $\mathfrak{L}$ be the pseudo-sphere $(x^1)^2 + (x^2)^2 - (x^3)^2 = -1$ for $x^3 > 0$ in $\mathbb{R}^3$ with the Lorentzian metric induced by the Minkowski metric $(dx^1)^2 + (dx^2)^2 - (dx^3)^2$.

The manifolds $\{\mathbb{E}^2, \mathfrak{H}, \mathbb{H}\}$ are Riemannian with sectional curvature $\{0, +1, -1\}$, respectively. The manifold $\mathfrak{L}$ is Lorentzian. There is another useful model for $\mathfrak{L}$. Let $\mathbb{L}$ be the upper half-plane $\mathbb{R}^+ \times \mathbb{R}$ with the metric $ds^2 = (x^1)^{-2}((dx^1)^2 - (dx^2)^2)$. This should be thought of as the Lorentzian hyperbolic plane. The manifolds $\{\mathbb{E}^2, \mathfrak{H}, \mathbb{H}, \mathfrak{L}\}$ are all geodesically complete. By contrast $\mathbb{L}$ is not geodesically complete. However, it is homogeneous and it embeds isometrically in $\mathfrak{L}$. We refer to D'Ascanio, Gilkey, and Pisani [22] for further details concerning the relationship between $\mathfrak{L}$ and $\mathbb{L}$.

In the pseudo-Riemannian category there are two levels of homogeneity. A locally homogeneous pseudo-Riemannian manifold is necessarily locally affine homogeneous, but there are pseudo-Riemannian manifolds which are not locally homogeneous but whose Levi-Civita connection is locally affine homogeneous (see Kowalski, Opozda, and Vlášek [43]).

12.4.2 2-DIMENSIONAL LIE GROUPS.

Give $\mathbb{R}^2$ the structure of an Abelian Lie group by using vector addition $(a^1, a^2) + (b^1, b^2) := (a^1 + b^1, a^2 + b^2)$. If we identify $\mathbb{R}^2$ with the translation group $T_{(a^1,a^2)} : (x^1, x^2) \rightarrow (x^1 + a^1, x^2 + a^2)$, then

$$T_{(a^1,a^2)+(b^1,b^2)} = T_{(a^1,a^2)} \circ T_{(b^1,b^2)}.$$

This Lie group can be identified with the additive group

$$\left\{ \begin{pmatrix} a_1 & 0 \\ 0 & a_2 \end{pmatrix} \text{ for } a^1 \in \mathbb{R}, \ a^2 \in \mathbb{R} \right\} \subset M_2(\mathbb{R}).$$

The multiplication $(a, b) * (c, d) := (ac, ad + b)$ makes $\mathbb{R}^+ \times \mathbb{R}$ into a non-Abelian Lie group. If we identify (a, b) with the affine map $A_{(a,b)} : x \to ax + b$, then $A_{(a,b)} \circ A_{(c,d)} = A_{(a,b)*(c,d)}$. Consequently, this non-Abelian Lie group is called the $ax + b$ group. We refer to the discussion in Section 6.7.2 of Book II for further details. It can be identified with the multiplicative group

$$\left\{ \begin{pmatrix} a & b \\ 0 & 1 \end{pmatrix} \text{ for } a > 0, \ b \in \mathbb{R} \right\} \subset \mathrm{GL}(2, \mathbb{R}).$$

Lemma 12.41

1. $\mathfrak{K}_A := \mathrm{span}\{\partial_{x^1}, \partial_{x^2}\}$ *is the Lie algebra of* $\mathbb{R}^2$.

2. $\mathfrak{K}_B := \mathrm{span}\{x^1 \partial_{x^1} + x^2 \partial_{x^2}, \partial_{x^2}\}$ *is the Lie algebra of the* $ax + b$ *group.*

3. *A connection on* $\mathbb{R}^2$ *is left-invariant if and only if* $\Gamma_{ij}{}^k \in \mathbb{R}$.

4. *A connection on the* $ax + b$ *group is left-invariant if and only if* $x^1 \Gamma_{ij}{}^k \in \mathbb{R}$.

5. *A simply connected 2-dimensional Lie group is isomorphic to one of these two groups.*

Proof. A direct computation establishes Assertions 1 and 2. If g belongs to a Lie group G, let L_g denote the action of G on itself by left-multiplication; an affine connection on G is *left-invariant* if and only if $L_g \circ \nabla = \nabla \circ L_g$, i.e., if left-multiplication is an affine map. Suppose $G = \mathbb{R}^2$. Let $(a_1, a_2) \in \mathbb{R}^2$. As $L_{(a_1, a_2)}$ preserves ∂_{x^1} and ∂_{x^2}, $L_{(a^1, a^2)}$ preserves ∇ if and only if

$$\Gamma_{ij}{}^k(x^1 + a^1, x^2 + a^2) = \Gamma_{ij}{}^k(x^1, x^2) \quad \text{for all} \quad (x^1, a^1, x^2, a^2) \in \mathbb{R}^4.$$

This implies that $\Gamma_{ij}{}^k \in \mathbb{R}$ and establishes Assertion 3. Let G be the $ax + b$ group. Then $L_{(a,b)}(x^1, x^2) = (ax^1, ax^2 + b)$ so $L_{(a,b)} \partial_{x^1} = a \partial_{x^1}$ and $L_{(a,b)} \partial_{x^2} = a \partial_{x^2}$. Thus ∇ is left-invariant if and only if $\Gamma_{ij}{}^k$ is independent of x^2 and $\Gamma_{ij}{}^k(ax^1) = a^{-1} \Gamma_{ij}{}^k(x^1)$; Assertion 4 follows. Let $\mathfrak{K}$ be the Lie algebra of a simply connected Lie group. If $\mathfrak{K}$ is Abelian, then $\mathfrak{K}$ is isomorphic to $\mathfrak{K}_A$ and if $\mathfrak{K}$ is non-Abelian, then $\mathfrak{K}$ is isomorphic to $\mathfrak{K}_B$; Assertion 5 now follows from this analysis. $\square$

Definition 12.42   Let $\mathcal{M} = (M, \nabla)$ be an affine surface, possibly with torsion.

1. We say that $\mathcal{M}$ is *Type A* if there exists a coordinate atlas for $\mathcal{M}$ so that $\Gamma_{ij}{}^k \in \mathbb{R}$.

2. We say that $\mathcal{M}$ is *Type B* if there exists a coordinate atlas for $\mathcal{M}$ so that $x^1 \Gamma_{ij}{}^k \in \mathbb{R}$.

3. We say that $\mathcal{M}$ is *Type C* if there exists a coordinate atlas for $\mathcal{M}$ so that ∇ is isomorphic to the Levi-Civita connection defined by the metric of the round sphere.

These structures have a profound affect on the geometry of the underlying affine structure. Let $\mathfrak{so}(3) := \mathrm{span}\{X, Y, Z\}$ be the 3-dimensional Lie algebra of the special orthogonal group. The bracket is given by $[X, Y] = Z$, $[Y, Z] = X$, and $[Z, X] = Y$. We refer to the discussion in Chapter 6 of Book II for further details.

Lemma 12.43   *Let* $\mathcal{M}$ *be a simply connected affine surface, possibly with torsion.*

1. *If M is Type A, Type B, or Type C, then M is locally affine homogeneous.*
2. *If M is Type A, there is an effective Lie subalgebra of $\mathfrak{K}(M)$ isomorphic to $\mathfrak{K}_A$.*
3. *If M is Type B, there is an effective Lie subalgebra of $\mathfrak{K}(M)$ isomorphic to $\mathfrak{K}_B$.*
4. *If M is Type C, there is an effective Lie subalgebra of $\mathfrak{K}(M)$ isomorphic to $\mathfrak{so}(3)$.*

Proof. It is clear from the discussion of Section 12.4.2 that any such geometry is locally affine homogeneous. Since M is simply connected, any local affine Killing vector field extends to a global affine Killing vector field. Assertion 1 and Assertion 2 follow from the proof of Lemma 12.41; the proof of Assertion 3 follows similar lines. □

12.4.3 THE LIE ALGEBRAS $\mathfrak{K}_A$, $\mathfrak{K}_B$, AND $\mathfrak{so}(3)$. We have the following converse to Lemma 12.43.

Lemma 12.44 *Let $M = (M, \nabla)$ be an affine surface, possibly with torsion. Let $\tilde{\mathfrak{K}}$ be an effective Lie subalgebra of $\mathfrak{K}(M)$.*

1. *If $\tilde{\mathfrak{K}}$ is isomorphic to $\mathfrak{K}_A$, then there exists a coordinate atlas for M so $\Gamma_{ij}{}^k \in \mathbb{R}$.*
2. *If $\tilde{\mathfrak{K}}$ is isomorphic to $\mathfrak{K}_B$, then there exists a coordinate atlas for M so $x^1 \Gamma_{ij}{}^k \in \mathbb{R}$.*
3. *If $\tilde{\mathfrak{K}}$ is isomorphic to $\mathfrak{so}(3)$, then there exists a coordinate atlas for M so ∇ is the Levi-Civita connection defined by the metric of the round sphere.*

Proof. Since M is assumed to be locally affine homogeneous, the analysis is local and we can work in an arbitrarily small neighborhood of the distinguished point P. We distinguish cases.

Suppose first that there exists an effective Lie subalgebra of $\mathfrak{K}(M)$ which is isomorphic to $\mathfrak{K}_A$. Choose affine Killing vector fields X and Y so that $X(P)$ and $Y(P)$ are linearly independent and so that $[X, Y] = 0$. By the Frobenius Theorem, there are local coordinates (x^1, x^2) centered at P on M so that $X = \partial_{x^1}$ and $Y = \partial_{x^2}$. It then follows that $\Gamma_{ij}{}^k \in \mathbb{R}$, which establishes Assertion 1.

Suppose next that there exists an effective Lie subalgebra of $\mathfrak{K}(M)$ which is isomorphic to $\mathfrak{K}_B$. Choose affine Killing vector fields X and Y so $X(P)$ and $Y(P)$ are linearly independent and so that $[X, Y] = aY$ for $a \neq 0$. We replace X by $a^{-1}X$ to assume that $[X, Y] = Y$. Choose local coordinates such that $Y = \partial_{x^2}$ and expand

$$X = u(x^1, x^2)\partial_{x^1} + v(x^1, x^2)\partial_{x^2}$$

for some functions u, v. The bracket relation $[X, Y] = Y$ shows $\partial_{x^2}u = 0$ and $\partial_{x^2}v = -1$. Consequently, $X = u(x^1)\partial_{x^1} + (v_0(x^1) - x^2)\partial_{x^2}$.

We change coordinates setting $\tilde{x}^1 = x^1$ and $\tilde{x}^2 = x^2 + \varepsilon(x^1)$. Then

$$d\tilde{x}^1 = dx^1, \qquad d\tilde{x}^2 = dx^2 + \varepsilon'(x^1)dx^1,$$
$$\partial_{\tilde{x}^1} = \partial_{x^1} - \varepsilon'(x^1)\partial_{x^2}, \quad \partial_{\tilde{x}^2} = \partial_{x^2}.$$

We then have $X = u(\tilde{x}^1)\partial_{\tilde{x}^1} + \{-\tilde{x}^2 + \varepsilon(\tilde{x}^1) + v_0(\tilde{x}^1) + u(\tilde{x}^1)\varepsilon'(\tilde{x}^1)\}\partial_{\tilde{x}^2}$. We may then solve the ODE $\varepsilon(x^1) + v_0(x^1) + u(x^1)\varepsilon'(x^1) = 0$ to express $X = u(\tilde{x}^1)\partial_{\tilde{x}^1} - \tilde{x}^2\partial_{\tilde{x}^2}$. Hence one may assume $X = u(x^1)\partial_{x^1} - x^2\partial_{x^2}$ without changing $Y = \partial_{x^2}$. Finally replace x^1 by $\hat{x}^1$ to ensure $u(x^1)\partial_{x^1} = -\hat{x}^1\partial_{\hat{x}^1}$ and $X = -\hat{x}^1\partial_{\hat{x}^1} - \hat{x}^2\partial_{\hat{x}^2}$. It now follows that the Christoffel symbols have the form given in Assertion 2.

Suppose finally that there is is an effective Lie subalgebra of $\mathfrak{K}(\mathcal{M})$ which is isomorphic to $\mathfrak{K}_B$. Choose affine Killing vector fields so $\{X, Y, Z\}$ is effective and so

$$[X, Y] = Z, \quad [Y, Z] = X, \quad [Z, X] = Y .$$

Since Z does not vanish identically, we may assume that $Z(P) \neq 0$ and choose local coordinates (x^1, x^2) which are defined near P so that $Z = \partial_{x^2}$. Decompose

$$X = u_1(x^1, x^2)\partial_{x^1} + \star\partial_{x^2}$$

where $\star$ indicates a coefficient which is not of interest. We then have $\partial^2_{x^2}u_1 = -u_1$ so $u_1(x^1, x^2) = r(x^1)\cos(x^2 + \theta(x^1))$. Since $\{X, Y, Z\}$ is a linearly independent set, $r(x^1) \neq 0$.

Change coordinates $(\tilde{x}^1, \tilde{x}^2) = (x^1, x^2 + \theta(x^1))$ so that $\partial_{\tilde{x}^2} = \partial_{x^2}$ and rewrite X in the form $X = r_1(\tilde{x}^1)\cos(\tilde{x}^2)\partial_{\tilde{x}^1} + \star\partial_{\tilde{x}^2}$ without changing ∂_{x^2}. We now choose coordinates $(z^1, z^2) = (f(\tilde{x}^1), \tilde{x}^2)$ so that $\partial_{z^1} = r_1(\tilde{x}^1)\partial_{\tilde{x}^1}$ and $\partial_{z^2} = \partial_{\tilde{x}^2}$. Consequently, $Z = \partial_{z^2}$. We use the Lie algebra relations defining $\mathfrak{so}(3)$ to expand X and Y in the form

$$X = \cos(z^2)\partial_{z^1} + \{v_c(z^1)\cos(z^2) + v_s(z^1)\sin(z^2)\}\partial_{z^2},$$
$$Y = -\sin(z^2)\partial_{z^1} + \{-v_c(z^1)\sin(z^2) + v_s(z^1)\cos(z^2)\}\partial_{z^2} .$$

To simplify the notation we replace (z^1, z^2) by (x^1, x^2). The bracket relation $[X, Y] = Z$ then yields the relations $-v_c(x^1) = 0$ and $-v_c(x^1)^2 - v_s(x^1)^2 + v_s'(x^1) = 1$. We solve this to obtain $v_c(x^1) = 0$ and $v_s(x^1) = \tan(x^1 + c)$. We replace $x^1 + c$ by x^1 to assume $c = 0$. We still have $Z = \partial_{x^2}$, but we now may express X and Y in a somewhat simpler form:

$$X = \cos(x^2)\partial_{x^1} + \tan(x^1)\sin(x^2)\partial_{x^2},$$
$$Y = -\sin(x^2)\partial_{x^1} + \tan(x^1)\cos(x^2)\partial_{x^2} .$$

Since $Z = \partial_{x^2}$ is an affine Killing vector field, the Christoffel symbols depend only on x^1. The affine Killing equations in general are quite complicated. For example, we have

$$\begin{aligned}
K_{11}{}^1 : 0 = & \cos(x^2)(\Gamma_{11}{}^1)'(x^1) + \Gamma_{11}{}^2(x^1)\sin(x^2) \\
& + \Gamma_{12}{}^1(x^1)\sec^2(x^1)\sin(x^2) + \Gamma_{21}{}^1(x^1)\sec^2(x^1)\sin(x^2) .
\end{aligned}$$

To simplify the equations, we set $x^2 = 0$ to obtain

$$\begin{aligned}
K_{11}{}^1 : & \ 0 = (\Gamma_{11}{}^1)', \\
K_{11}{}^2 : & \ 0 = -\Gamma_{11}{}^2\tan(x^1) + (\Gamma_{11}{}^2)', \\
K_{12}{}^1 : & \ 0 = \Gamma_{12}{}^1\tan(x^1) + (\Gamma_{12}{}^1)', \\
K_{12}{}^2 : & \ 0 = \sec^2(x^1) + (\Gamma_{12}{}^2)',
\end{aligned} \qquad (12.4.\text{a})$$

and

$$K_{21}{}^1 : \; 0 = \Gamma_{21}{}^1 \tan(x^1) + (\Gamma_{21}{}^1)',$$
$$K_{21}{}^2 : \; 0 = \sec^2(x^1) + (\Gamma_{21}{}^2)',$$
$$K_{22}{}^1 : \; 0 = -1 + 2\Gamma_{22}{}^1 \tan(x^1) + (\Gamma_{22}{}^1)', \tag{12.4.b}$$
$$K_{22}{}^2 : \; 0 = \Gamma_{22}{}^2 \tan(x^1) + (\Gamma_{22}{}^2)'.$$

Let $a_{ij}{}^k$ be constants to be determined. We solve the ODEs given in Equation (12.4.a) and in Equation (12.4.b) to see:

$$\Gamma_{11}{}^1 = a_{11}{}^1, \qquad\qquad\qquad \Gamma_{11}{}^2 = a_{11}{}^2 \sec(x^1),$$
$$\Gamma_{12}{}^1 = a_{12}{}^1 \cos(x^1), \qquad\qquad \Gamma_{12}{}^2 = a_{12}{}^2 - \tan(x^1),$$
$$\Gamma_{21}{}^1 = a_{21}{}^1 \cos(x^1), \qquad\qquad \Gamma_{21}{}^2 = a_{21}{}^2 - \tan(x^1),$$
$$\Gamma_{22}{}^1 = a_{22}{}^1 \cos(x^1)^2 + \cos(x^1)\sin(x^1), \quad \Gamma_{22}{}^2 = a_{22}{}^2 \cos(x^1).$$

Equation (12.4.a) and Equation (12.4.b) were obtained by specializing the affine Killing equations at $x^2 = 0$. We now specialize the affine Killing equations at $x^2 = \frac{\pi}{2}$ to determine the constants. Equation (12.4.a) becomes:

$$K_{11}{}^1 : \; 0 = (a_{11}{}^2 + a_{12}{}^1 + a_{21}{}^1)\sec(x^1),$$
$$K_{11}{}^2 : \; 0 - (-a_{11}{}^1 + a_{12}{}^2 + a_{21}{}^2)\sec(x^1)^2,$$
$$K_{12}{}^1 : \; 0 = -a_{11}{}^1 + a_{12}{}^2 + a_{22}{}^1,$$
$$K_{12}{}^2 : \; 0 = -(a_{11}{}^2 + a_{12}{}^1 - a_{22}{}^2)\sec(x^1)$$

which leads to the relations.

$$a_{21}{}^1 = -a_{11}{}^2 - a_{12}{}^1, \quad a_{21}{}^2 = a_{11}{}^1 - a_{12}{}^2,$$
$$a_{22}{}^1 = a_{11}{}^1 - a_{12}{}^2, \qquad a_{22}{}^2 = a_{11}{}^2 + a_{12}{}^1.$$

We specialize the affine Killing equations $K_{21}{}^1$ and $K_{22}{}^2$ at $x^2 = \frac{\pi}{2}$ to obtain

$$K_{21}{}^1 : \; 0 = a_{11}{}^1 - 2a_{12}{}^2, \quad K_{22}{}^2 : \; 0 = -2a_{11}{}^1 + a_{12}{}^2.$$

This yields that $a_{11}{}^1 = 0$ and $a_{12}{}^2 = 0$. The remaining affine Killing equations without specialization yield

$$(2a_{11}{}^2 + a_{12}{}^1)\cos(x^1)\sin(x^2) = 0 \quad \text{and} \quad (a_{11}{}^2 + 2a_{12}{}^1)\sec(x^1)\sin(x^2) = 0$$

so $a_{11}{}^2 = 0$ and $a_{12}{}^1 = 0$. This yields finally

$$\Gamma_{11}{}^1 = 0, \quad \Gamma_{11}{}^2 = 0, \qquad\quad \Gamma_{12}{}^1 = 0, \qquad\qquad \Gamma_{12}{}^2 = -\tan(x^1),$$
$$\Gamma_{21}{}^1 = 0, \quad \Gamma_{21}{}^2 = -\tan(x^1), \quad \Gamma_{22}{}^1 = \cos(x^1)\sin(x^1), \quad \Gamma_{22}{}^2 = 0.$$

We summarize matters. If we assume $\mathfrak{so}(3)$ is an effective Lie subalgebra of $\mathfrak{K}(M)$, then the Christoffel symbols are uniquely determined. Thus any two such geometries are locally isomorphic. Since $\mathfrak{so}(3)$ is the Lie algebra of SO(3) and since SO(3) is the Lie group of orientation-preserving isometries of S^3, we obtain that any such geometry is locally modeled on the geometry of the round sphere.   □

Remark 12.45   One may show directly that there are no additional affine Killing vector fields for the round sphere and thus $\mathfrak{so}(3) = \mathfrak{K}(M)$. Consequently, no Type C geometry is either Type A or Type B. Any flat geometry is trivially both Type A and Type B. However, as we shall see in Section 14.2.4, there are locally affine homogeneous surfaces which are not flat and which are both Type A and Type B; they admit both an effective Lie subalgebra isomorphic to $\mathfrak{K}_A$ and an effective Lie subalgebra isomorphic to $\mathfrak{K}_B$.

12.4.4 THE LIE ALGEBRA $A^a_{3,7}$.   We shall adopt the notation of Patera et al. [55] and let $A^a_{3,7} = \mathrm{span}\{X, Y\}$ be the 3-dimensional Lie algebra defined by the bracket relations $[X, Y] = 0$, $[X, Z] = aX + Y$, and $[Y, Z] = aY - X$. Note that $A^a_{3,7}$ is a semi-direct product $\mathbb{R} \ltimes_\varphi \mathbb{R}^2$ where $\mathbb{R}Z$ acts on $\mathbb{R}^2 = \mathrm{span}\{X, Y\}$ by the derivation

$$\varphi = \begin{pmatrix} -a & 1 \\ -1 & -a \end{pmatrix}.$$

The Lie algebra $A^a_{3,7}$ is a Lie subalgebra of $\mathfrak{K}(M)$ for the Type A structures $\mathcal{M}^4_5(c)$ (see Definition 13.7). The following result shows that $A^a_{3,7}$ does not affect any classification results.

Lemma 12.46   *Suppose that there is an effective Lie subalgebra of $\mathfrak{K}(M)$ which is isomorphic to $A^a_{3,7}$. Then there exists an effective Lie subalgebra of $\mathfrak{K}(M)$ isomorphic to $\mathfrak{K}_A$.*

Proof.   The result is immediate if $\{X, Y\}$ is effective. Consequently, we assume that Y is a multiple of X and $\{X, Z\}$ is effective. We use Lemma 12.4 to normalize the coordinate system so that

$$X = \partial_{x^2} \text{ and } Y = v(x^1, x^2)\partial_{x^2}.$$

Since $[X, Y] = 0$, $\partial_{x^2}v = 0$. Thus, $v = v(x^1)$. Since Y is not a constant multiple of X, we may assume that $v'(x^1) \neq 0$. Therefore, we have that $X = \partial_{x^2}$ and $Y = v(x^1)\partial_{x^2}$. Consequently, by Assertion 1 of Lemma 12.5, v solves the ODE $(\Gamma_{12}{}^2 + \Gamma_{21}{}^2)v' + v'' = 0$. Expand

$$\begin{aligned} Z &= u(x^1, x^2)\partial_{x^1} + w(x^1, x^2)\partial_{x^2}, \\ [X, Z] &= \partial_{x^2}u(x^1, x^2)\partial_{x^1} + \partial_{x^2}w(x^1, x^2)\partial_{x^2} \\ &= aX + Y = (a + v(x^1))\partial_{x^2}. \end{aligned}$$

Thus $u = u(x^1)$ and $w = (a + v(x^1))x^2 + v_0(x^1)$; as $\{X, Z\}$ is effective, $u \neq 0$ and

$$X = \partial_{x^2}, \quad Y = v(x^1)\partial_{x^2}, \quad Z = u(x^1)\partial_{x^1} + \{(a + v(x^1))x^2 + v_0(x^1)\}\partial_{x^2}.$$

The affine Killing equation $K_{11}{}^2$ for Z yields $0 = (\Gamma_{12}{}^2 + \Gamma_{21}{}^2)v_0' + v_0''$. Thus v and v_0 are solutions of the same linear homogeneous ODE. Since $\{\mathbb{1}, v\}$ are linearly independent solutions, they form a basis for the solution space. Thus after subtracting a suitable linear combination of ∂_{x^2} and $v(x^1)\partial_{x^2}$, we may assume $v_0 = 0$ so $Z = u(x^1)\partial_{x^1} + (a + v(x^1))x^2\partial_{x^2}$. The relations

$$\Gamma_{11}{}^1 = 0, \quad \Gamma_{11}{}^2 = 0, \quad \Gamma_{12}{}^1 = 0, \quad \Gamma_{21}{}^1 = 0,$$
$$\Gamma_{22}{}^1 = 0, \quad \Gamma_{22}{}^2 = 0, \quad (\Gamma_{12}{}^2 + \Gamma_{21}{}^2)v' + v'' = 0$$

in Lemma 12.5 1 show that $x^2\partial_{x^2}$ is an affine Killing vector field. Since $x^2\partial_{x^2}$ commutes with Z, we obtain an effective Lie subalgebra of $\mathfrak{K}(\mathcal{M})$ isomorphic to $\mathfrak{K}_A$. $\qquad\square$

12.4.5 THE ADJOINT ACTION ON $\mathfrak{K}(\mathcal{M})$. Let X be an affine Killing vector field with $X(P) \neq 0$. We impose the normalizations of Lemma 12.4 to assume

$$X = \partial_{x^2}, \quad \Gamma_{ij}{}^k(x^1, x^2) = \Gamma_{ij}{}^k(x^1), \quad \Gamma_{11}{}^1(x^1) = 0, \quad \Gamma_{11}{}^2(x^1) = 0.$$

Let $\mathrm{ad}(\partial_{x^2})(Y) := [X, Y]$ denote the adjoint action of ∂_{x^2} on $\mathfrak{K}_{\mathbb{C}}(\mathcal{M}) := \mathfrak{K}(\mathcal{M}) \otimes_{\mathbb{R}} \mathbb{C}$. Denote the generalized eigenspaces of this action by

$$\mathfrak{E}(\alpha) := \{X_\alpha \in \mathfrak{K}_{\mathbb{C}}(\mathcal{M}) : (\mathrm{ad}(\partial_{x^2}) - \alpha)^6 X_\alpha = 0\}.$$

Choose $X \in \mathfrak{E}(\alpha)$ for some α so $\{X, \partial_{x^2}\}$ is effective. Expand

$$X = e^{\alpha x^2} \sum_{i=0}^{i_0} u_i(x^1)(x^2)^i \partial_{x^1} + e^{\alpha x^2} \sum_{j=0}^{j_0} v_j(x^1)(x^2)^j \partial_{x^2}.$$

Since $\{X, \partial_{x^2}\}$ is effective, $u_i \neq 0$ for some i. Choose i_0 maximal so $u_{i_0} \neq 0$. By applying $(\mathrm{ad}(\partial_{x^2}) - \alpha)^{i_0-1}$ to X, we may assume that $i_0 = 0$ so

$$X = e^{\alpha x^2}\{u(x^1)\partial_{x^1} + \sum_{j=0}^{j_0} v_j(x^1)(x^2)^j \partial_{x^2}\} \quad \text{for} \quad u \neq 0. \tag{12.4.c}$$

We first examine $\mathfrak{E}(\alpha)$ for $\alpha \neq 0$. The case $\alpha = 0$ will be considered in Lemma 12.48.

Lemma 12.47 *If $\alpha \neq 0$, then there exists an effective Lie subalgebra of $\mathfrak{K}(\mathcal{M})$ isomorphic to $\mathfrak{K}_A$, $\mathfrak{K}_B$, or $\mathfrak{so}(3)$.*

Proof. Adopt the notation established above. We wish to show $j_0 = 0$. Suppose to the contrary that $v_j \neq 0$ for some $j > 0$. Choose v_{j_0} maximal so $v_{j_0} \neq 0$. Hence

$$0 \neq (\mathrm{ad}(\partial_{x^2}) - \alpha)^{j_0} X = j_0! e^{\alpha x^2} v_{j_0}(x^1)\partial_{x^2} \in \mathfrak{K}_\alpha(\mathcal{M}).$$

Lemma 12.5 2 now implies $u(x^1) = 0$ contrary to our assumption. Thus $j_0 = 0$. We consider subsequently the different possibilities for α to be real or complex.

Case 1. Suppose $\alpha \in \mathbb{R}$. Since $X = e^{\alpha x^2}\{u(x^1)\partial_{x^1} + v(x^1)\partial_{x^2}\}$, one has $[\partial_{x^2}, X] = \alpha X$ so $[\alpha^{-1}\partial_{x^2}, X] = X$. Since $\{X, \partial_{x^2}\}$ is effective, we have an effective Lie subalgebra isomorphic to $\mathfrak{K}_\mathcal{B}$.

Case 2. Suppose $\alpha \in \mathbb{C} \setminus \mathbb{R}$. By rescaling x^2, we may suppose $\alpha = a + \sqrt{-1}$ for $a \geq 0$. Now we consider the following possibilities.

Case 2.1. Assume that the real part $a \neq 0$. Choose a maximal so that there exists X in $\mathfrak{E}(a + \sqrt{-1})$ so $\{X, \partial_{x^2}\}$ is effective. Observe that $[\mathfrak{E}(\alpha), \mathfrak{E}(\beta)] \subset \mathfrak{E}(\alpha + \beta)$ for arbitrary complex numbers α and β. Indeed, $X_\alpha \in \mathfrak{E}(\alpha)$ and $X_\beta \in \mathfrak{E}(\beta)$ if and only if

$$X_\alpha = e^{\alpha x^2} \sum_i u_i^\alpha(x^1)(x^2)^i \partial_{x^1} + e^{\alpha x^2} \sum_j v_j^\alpha(x^1)(x^2)^j \partial_{x^2},$$

$$X_\beta = e^{\beta x^2} \sum_k \tilde{u}_k^\beta(x^1)(x^2)^k \partial_{x^1} + e^{\beta x^2} \sum_\ell \tilde{v}_\ell^\beta(x^1)(x^2)^\ell \partial_{x^2}.$$

This leads to an expansion for $[X_\alpha, X_\beta]$ where the relevant exponential is $e^{(\alpha+\beta)x^2}$ that shows $[X_\alpha, X_\beta] \in \mathfrak{E}(\alpha + \beta)$.

Expand $X \in \mathfrak{E}(a + \sqrt{-1})$ as $X = e^{ax^2} e^{\sqrt{-1}x^2}\{u(x^1)\partial_{x^1} + v(x^1)\partial_{x^2}\}$. We have now that $\bar{X} \in \mathfrak{E}(\bar{\alpha})$. Let $Y_1 := \sqrt{-1}[X, \bar{X}]$. We showed previously that

$$Y_1 \in \mathfrak{E}(\alpha + \bar{\alpha}) = \mathfrak{E}(2a).$$

Since $\bar{Y}_1 = Y_1$, Y_1 is real. Decompose

$$Y_1 = e^{2ax^2}\{u_1(x^1)\partial_{x^1} + v_1(x^1)\partial_{x^2}\}.$$

Case 2.1.1. If $u_1 \neq 0$, then we may apply Case 1 to Y_1.

Case 2.1.2. If $u_1 = 0$ and if $v_1 \neq 0$, then $\Gamma_{22}{}^2 = -2a$ by Lemma 12.5 2. Set

$$Y_2 := [X, Y_1] = e^{(3a+\sqrt{-1})x^2}\{u_2(x^1)\partial_{x^1} + v_2(x^1)\partial_{x^2}\} \in \mathfrak{E}(3a + \sqrt{-1}).$$

Then one of the following three possibilities hold.

Case 2.1.2.a. If $u_2 \neq 0$, this contradicts the maximality of a.

Case 2.1.2.b. If $u_2 = 0$ and $v_2 \neq 0$, then by Lemma 12.5 2 with $\Gamma_{22}{}^2(x^1) = -(3a + \sqrt{-1})$. This contradicts the fact that $\Gamma_{22}{}^2(x^1) = -2a$.

Case 2.1.2.c. Finally, if $u_2 = 0$ and $v_2 = 0$, then X and Y_1 commute. Since Y_1 is real, $[\Re(X), Y_1] = 0$ and $[\Im(X), Y_1] = 0$. Either $\{\Re(X), Y_1\}$ or $\{\Im(X), Y_1\}$ generates an effective 2-dimensional Lie subalgebra of $\mathfrak{K}(\mathcal{M})$ which is isomorphic to $\mathfrak{K}_\mathcal{A}$.

Case 2.1.3. If $u_1 = 0$ and $v_1 = 0$, then $[X, \bar{X}] = 0$ and $\{\Re(X), \Im(X), \partial_{x^2}\}$ span a Lie algebra

$$[\Im(X), \Re(X)] = 0, \quad [\Im(X), -\partial_{x^2}] = a\Im(X) + \Re(X), \quad [\Re(X), -\partial_{x^2}] = a\Re(X) - \Im(X)$$

isomorphic to $A_{3,7}^a$. Hence there exists an effective Lie subalgebra of $\mathfrak{K}(\mathcal{M})$ isomorphic to $\mathfrak{K}_\mathcal{A}$ by Lemma 12.46.

Case 2.2. Assume that α is purely imaginary $\alpha = \sqrt{-1}$. We have X_i in $\mathfrak{K}(\mathcal{M})$ with $\{X_i, \partial_{x^2}\}$ effective where

$$X_1 = u(x^1, x^2)\partial_{x^1} + v(x^1, x^2)\partial_{x^2}, \quad X_2 = \partial_{x^2} X_1,$$
$$u(x^1, x^2) = u_1(x^1)\cos(x^2) + u_2(x^1)\sin(x^2),$$
$$v(x^1, x^2) = v_1(x^1)\cos(x^2) + v_2(x^1)\sin(x^2).$$

Let $X_3 := [X_1, X_2] \in \mathfrak{E}(0)$. There are no polynomial terms in X_1 or X_2. Consequently,

$$X_3 = u_3(x^1)\partial_{x^1} + v_3(x^1)\partial_{x^2}$$

and one of the following possibilities pertain.

Case 2.2.1. If $u_3 \neq 0$, then $\{X_3, \partial_{x^2}\}$ is an effective Lie algebra isomorphic to $\mathfrak{K}_A$.

Case 2.2.2. If $u_3 = 0$ but $v_3 \neq 0$, then $X_3 = v_3(x^1)\partial_{x^2}$ and one of the following two possibilities occurs.

Case 2.2.2.a. If $v_3' \neq 0$, then Assertion 2 in Lemma 12.5 1 now gives $u_1 = u_2 = 0$, which is false.

Case 2.2.2.b. Suppose $v_3' = 0$ so $[X_1, X_2]$ is a constant non-zero multiple of ∂_{x^2}. This gives the Lie algebra $\mathfrak{so}(3)$.

Case 2.2.3. If $X_3 = 0$, we have $[X_1, X_2] = 0$. Then $\{X = X_1, Y = -X_2, Z = \partial_{x^2}\}$ span the Lie algebra $A_{3,7}^0$ and we can apply Lemma 12.46.   □

We finally examine $\mathfrak{E}(0)$.

Lemma 12.48   *Assume that $\alpha = 0$ and that there exists $X \in \mathfrak{E}(0)$ such that $\{X, \partial_{x^2}\}$ is effective. Then there exists an effective Lie subalgebra of $\mathfrak{K}(\mathcal{M})$ isomorphic to $\mathfrak{K}_A$, $\mathfrak{K}_B$, or $\mathfrak{so}(3)$.*

Proof. Choose $X \in \mathfrak{E}(0)$ of the form given in Equation (12.4.c), i.e.,

$$X = e^{\alpha x^2}\{u(x^1)\partial_{x^1} + \sum_{j=0}^{j_0} v_j(x^1)(x^2)^j \partial_{x^2}\} \quad \text{for} \quad u \neq 0.$$

If $j_0 = 0$, then $\{X, \partial_{x^2}\}$ is an effective algebra isomorphic to $\mathfrak{K}_A$. We may therefore assume that $j_0 \geq 1$. We suppose $j_0 \geq 2$ and argue for a contradiction. Since $j_0 - 1 \leq 2j_0 - 3$, $u(x^1)\partial_{x^2}$ contributes lower-order terms and plays no role. Set:

$$Y_1 := [\partial_{x^2}, X] = \{c_1 v_{j_0}(x^1)(x^2)^{j_0-1} + O((x^2)^{j_0-2})\}\partial_{x^2},$$
$$Y_2 := [X, Y_1] = \{c_2 v_{j_0}^2(x^1)(x^2)^{2(j_0-1)} + O((x^2)^{2(j_0-1)-1})\}\partial_{x^2},$$
$$\cdots \qquad \cdots$$
$$Y_n := [X, Y_{n-1}] = \{c_n v_{j_0}^n(x^1)(x^2)^{n(j_0-1)} + O((x^2)^{n(j_0-1)-1})\}\partial_{x^2}$$

where the constants $c_n \neq 0$. This creates an infinite string of linearly independent elements of $\mathfrak{K}(\mathcal{M})$ which is not possible. We therefore suppose $j_0 = 1$ henceforth so

$$X = u(x^1)\partial_1 + (v_1(x^1)x^2 + v_0(x^1))\partial_{x^2} \quad \text{for} \quad v_1 \neq 0\,.$$

If $v_1' = 0$, then $[X, \partial_{x^2}] = v_1\partial_{x^2}$ and we obtain a subalgebra isomorphic to $\mathfrak{K}_A$ or $\mathfrak{K}_B$. We therefore suppose $v' \neq 0$ and apply Lemma 12.5 1 to obtain the relations of Assertion 1. If $w(x^1)\partial_{x^1} \in \mathfrak{K}(\mathcal{M})$, we obtain an affine Killing equation

$$K_{11}{}^2 : \ 0 = (\Gamma_{12}{}^2 + \Gamma_{21}{}^2)w' + w'' = 0\,.$$

This is a linear homogeneous second-order ODE. Since both $v_1(x^1)$ and $\mathbb{1}$ satisfy this ODE, w is a linear combination of v_1 and $\mathbb{1}$. Let $\mathfrak{K}_0 := \mathrm{span}\{v(x^1)\partial_{x^2}, \partial_{x^2}\}$; $\mathrm{ad}(X)$ preserves this space. We now change our perspective and decompose $\mathfrak{K}_0 \otimes \mathbb{C}$ into generalized eigenspaces under the action of $\mathrm{ad}(X)$. If there is an eigenvalue β with $\beta \neq 0$, the analysis of Lemma 12.47 pertains since $\{X, \xi\}$ is an effective set for any $0 \neq \xi \in \mathfrak{K}_0$. On the other hand if 0 is an eigenvalue, there exists $\xi \in \mathfrak{K}_0$ so that $[X, \xi] = 0$ and there exists an effective Lie subalgebra which is isomorphic to $\mathfrak{K}_A$. $\square$

12.4.6 CLASSIFICATION OF AFFINE HOMOGENEOUS SURFACES. The classification of affine homogeneous surfaces by Opozda [53] (see also Arias-Marco and Kowalski [2]) can now be stated in terms of the existence of distinguished coordinates as follows.

Theorem 12.49   *Let M be a smooth connected manifold and let ∇ be a connection, possibly with torsion, on the tangent bundle of M.*

1. *If $\mathcal{M} = (M, \nabla)$ is locally affine homogeneous, then there exists an effective Lie subalgebra $\tilde{\mathfrak{K}}$ of $\mathfrak{K}_P(\mathcal{M})$ which is isomorphic to $\mathfrak{K}_A$, to $\mathfrak{K}_B$, or to $\mathfrak{so}(3)$.*

   (a) If $\tilde{\mathfrak{K}} \approx \mathfrak{K}_A$, then there exists a coordinate atlas so $\Gamma_{ij}{}^k \in \mathbb{R}$.

   (b) If $\tilde{\mathfrak{K}} \approx \mathfrak{K}_B$, then there exists a coordinate atlas so $\Gamma_{ij}{}^k = (x^1)^{-1}A_{ij}{}^k$ for $A_{ij}{}^k \in \mathbb{R}$.

   (c) If $\tilde{\mathfrak{K}} \approx \mathfrak{so}(3)$, then there exists a coordinate atlas so ∇ is the Levi-Civita connection defined by the metric of the round sphere.

2. *If there exists an effective Lie subalgebra of $\mathfrak{K}(\mathcal{M})$ which is isomorphic to $\mathfrak{K}_A$, $\mathfrak{K}_B$, or $\mathfrak{so}(3)$, then $\mathcal{M}$ is locally affine homogeneous.*

3. *If there exists a coordinate atlas for $\mathcal{M}$ normalized as in Assertions 1-a, 1-b or 1-c, then $\mathcal{M}$ is locally affine homogeneous.*

Proof. The proof of Theorem 12.49 now follows from the previous lemmas. Lemma 12.47 and Lemma 12.48 show the existence of an effective Lie subalgebra $\mathfrak{K}_0 \subset \mathfrak{K}(\mathcal{M})$ isomorphic to $\mathfrak{K}_A$, $\mathfrak{K}_B$ or $\mathfrak{so}(3)$. The existence of distinguished coordinates then follows from Lemma 12.44. The remaining assertions follow from Lemma 12.41 and Lemma 12.43. $\square$

We note, in passing, that recent work of D'Ascanio et al. [24] completes the analysis of Assertion 1 by determining exactly which Lie algebras arise, up to isomorphism, as the Lie algebras of affine Killing vector fields of locally affine homogeneous simply connected manifolds; this is implicit, of course, in the work of Arias-Marco and Kowalski [2].

12.4.7 METRIZABILITY. If (M, g) is a pseudo-Riemannian manifold, then we can obtain an affine manifold by taking ∇ to be the Levi-Civita connection; such an affine structure is said to be *metrizable*. There are, however, many affine manifolds which are not metrizable. For example, the Ricci tensor of any pseudo-Riemannian manifold is symmetric; in Chapter 14, we will discuss structures where this fails and which therefore are not metrizable.

The situation is more tractable in the 2-dimensional case since the Ricci tensor of (M, g) satisfies $\rho = \frac{1}{2} {}^g\text{Sc}\, g$, where ${}^g\text{Sc}$ denotes the scalar curvature of (M, g). Hence the Ricci tensor is either parallel or recurrent since $\nabla\rho = d\, {}^g\text{Sc} \otimes \rho$ (see Vanžurová [60]).

Locally homogeneous metrizable affine connections on surfaces were determined by Kowalski, Opozda, and Vlášek [43] as follows.

Theorem 12.50 *The only 2-dimensional locally non–homogeneous pseudo–Riemannian metrics with locally homogeneous Levi-Civita connection are those which are modeled on $\mathcal{N}_\varepsilon := (\mathbb{R}^+ \times \mathbb{R}, g_\varepsilon^k)$, where $g_\varepsilon^k = (x^1)^{2k}(dx^1 \otimes dx^1 + \varepsilon\, dx^2 \otimes dx^2)$ for $\varepsilon = \pm 1$ and $k \neq 0, -1$.*

The corresponding connections, which are of Type $\mathcal{B}$, are completely determined by ${}^\varepsilon\Gamma_{ij}{}^k = (x^1)^{-1}\, {}^\varepsilon A_{ij}{}^k$, where

$$ {}^\varepsilon A_{11}{}^1 = k, \; {}^\varepsilon A_{11}{}^2 = 0, \; {}^\varepsilon A_{12}{}^1 = 0, \; {}^\varepsilon A_{12}{}^2 = k, \; {}^\varepsilon A_{22}{}^1 = -\varepsilon k, \; {}^\varepsilon A_{22}{}^2 = 0. $$

It is now a routine calculation to show that $\rho_\varepsilon = \frac{k}{(x^1)^2}(dx^1 \otimes dx^1 + \varepsilon\, dx^2 \otimes dx^2)$ is recurrent, symmetric, and of rank 2. Furthermore, $\dim(\mathfrak{K}(\mathcal{N}_\varepsilon)) = 2$ and moreover $f(x^1) = (x^1)^{1+2k}$ is a solution of the affine quasi-Einstein equation for $\mu = 1 + 2k$.

12.5 ANALYTIC STRUCTURE FOR HOMOGENEOUS AFFINE SURFACES

In Section 12.5.1, we examine the structure group of a Type $\mathcal{A}$ or Type $\mathcal{B}$ coordinate atlas. We use this analysis in Section 12.5.2 to show such an atlas is real analytic and thereby, in light of Theorem 12.49, give a natural real analytic structure to any locally affine homogeneous surface. In Section 12.5.3, we examine the module structure of various tensors under the action of the Lie algebras $\mathfrak{K}_\mathcal{A}$ and $\mathfrak{K}_\mathcal{B}$ of the two 2-dimensional Lie groups.

12.5.1 COORDINATE ATLAS. The general linear group $\text{GL}(2, \mathbb{R})$ acts by matrix multiplication on $\mathbb{R}^2$; we say that two connections on $\mathbb{R}^2$ are linearly equivalent if they are intertwined by an element of $\text{GL}(2, \mathbb{R})$. We say that two connections on $\mathbb{R}^+ \times \mathbb{R}$ are linearly equivalent if they are intertwined by a linear transformation $(x^1, x^2) \to (x^1, ux^1 + vx^2)$ for $u \in \mathbb{R}$ and $v \neq 0$.

Theorem 12.51   *Let M be a locally affine homogeneous surface with $\dim(\mathfrak{K}_P(M)) = 2$ for all points P in M. If M admits a Type $\mathcal{A}$ atlas, then the coordinate transformations have the form $\vec{x} \to A\vec{x} + \vec{b}$ for $A \in \mathrm{GL}(2, \mathbb{R})$ and $\vec{b} \in \mathbb{R}^2$. If M admits a Type $\mathcal{B}$ atlas, then coordinate transformations have the form $(x^1, x^2) \to (ax^1, bx^1 + cx^2 + d)$ for $a > 0$ and $c \neq 0$.*

Proof.   Let $(\mathcal{O}_\alpha, \phi_\alpha)$ be a Type $\mathcal{A}$ coordinate atlas. Let ∇_α be the associated Type $\mathcal{A}$ connections on $\mathcal{O}_\alpha$ and let $\mathfrak{K}_\alpha$ be the associated Lie algebras of affine Killing vector fields. Then the transition functions $\phi_{\alpha\beta}$ intertwine ∇_α and ∇_β and, consequently, intertwine the Lie algebras $\mathfrak{K}_\alpha$ and $\mathfrak{K}_\beta$. Suppose that $\dim(\mathfrak{K}_\alpha) = 2$. We then have $\mathfrak{K}_\alpha = \mathrm{span}\{\partial_{x_\alpha^1}, \partial_{x_\alpha^2}\}$. Consequently,

$$(\phi_{\alpha\beta})_* \partial_{x_\alpha^i} = a_i^j \partial_{x_\beta^j} \text{ for } (a_i^j) \in \mathrm{GL}(2, \mathbb{R}).$$

If we express

$$\phi_{\alpha\beta}(x^1, x^2) = (\phi_{\alpha\beta}^1(x^1, x^2), \phi_{\alpha\beta}^2(x^1, x^2)),$$

then $\partial_{x^i}\phi_{\alpha\beta}^j = a_i^j$ is constant so $\phi_{\alpha\beta}$ has the required form.

Suppose the atlas is a Type $\mathcal{B}$ atlas. Let $X := x^1 \partial_{x^1} + x^2 \partial_{x^2}$. Since the left-action of the $ax + b$ group defines a transitive action on $\mathbb{R}^+ \times \mathbb{R}$, we may assume without loss of generality that the transition functions $\phi_{\alpha\beta}$ satisfy $\phi_{\alpha\beta}(1, 0) = (1, 0)$. Suppose $\dim(\mathfrak{K}(M)) = 2$. Expand $\phi_{\alpha\beta} = (\phi_{\alpha\beta}^1, \phi_{\alpha\beta}^2)$. We must show $\phi_{\alpha\beta}^1 = x^1$ and $\phi_{\alpha\beta}^2 = bx^1 + cx^2$ for $c \neq 0$. Because $\mathfrak{K}_\alpha = \mathrm{span}\{X, \partial_{x^2}\}$, $(\phi_{\alpha\beta})_* X = rX + s\partial_{x^2}$ and $(\phi_{\alpha\beta})_* \partial_{x^2} = tX + v\partial_{x^2}$ for suitably chosen constants. We have that $\partial_{x^2} = [\partial_{x^2}, X]$ and thus

$$tX + v\partial_{x^2} = [tX + v\partial_{x^2}, rX + s\partial_{x^2}] = (vr - ts)\partial_{x^2}.$$

This implies that $t = 0$ and $r = 1$ so $(\phi_{\alpha\beta})_* X = X + s\partial_{x^2}$ and $(\phi_{\alpha\beta})_* \partial_{x^2} = v\partial_{x^2}$. Let $\tilde{\phi}(x^1, x^2) = (x^1, \alpha x^1 + \beta x^2 - \alpha)$. Then $\tilde{\phi}(1, 0) = (1, 0)$, and

$$\tilde{\phi}^*(dx^1) = dx^1, \qquad\qquad \tilde{\phi}^*(dx^2) = \alpha dx^1 + \beta dx^2,$$
$$\tilde{\phi}_*(\partial_{x^1}) = \partial_{x^1} - \alpha\beta^{-1}\partial_{x^2}, \qquad\qquad \tilde{\phi}_*(\partial_{x^2}) = \beta^{-1}\partial_{x^2},$$
$$\tilde{\phi}_*(x^1\partial_{x^1} + x^2\partial_{x^2}) = x^1(\partial_{x^1} - \alpha\beta^{-1}\partial_{x^2}) + (\alpha x^1 + \beta x^2 - \alpha)(\beta^{-1}\partial_{x^2})$$
$$= (x^1\partial_{x^1} + x^2\partial_{x^2}) - \alpha\beta^{-1}\partial_{x^2}.$$

To ensure $(\tilde{\phi} - \phi_{\alpha\beta})_* = 0$, we take $-\alpha\beta^{-1} = s$ and $\beta^{-1} = v$. We then have $\tilde{\phi}_* - \phi_{\alpha\beta}$ is constant. Since $(\tilde{\phi} - \phi_{\alpha\beta})(1, 0) = (0, 0)$, we conclude $\tilde{\phi} = \phi_{\alpha\beta}$ as desired. $\square$

12.5.2 REAL ANALYTIC STRUCTURES.

Theorem 12.52   *If M is a Type $\mathcal{A}$ (resp. Type $\mathcal{B}$) surface, then the Type $\mathcal{A}$ (resp. Type $\mathcal{B}$) coordinate atlas is real analytic.*

Proof. Suppose first that $\mathcal{M}$ has a Type $\mathcal{A}$ model. Let $\{\mathcal{U}_\alpha, \tilde{x}_\alpha\}$ be the associated Type $\mathcal{A}$ coordinate atlas. The transition map $\Phi_{\alpha\beta}$ from an open subset of $\mathcal{U}_\alpha$ to an open subset of $\mathcal{U}_\beta$ is an affine map. We must show $\Phi_{\alpha\beta}$ is real analytic. The vector fields $\{\partial_{x_\beta^1}, \partial_{x_\beta^2}\}$ are affine Killing vector fields. The dual frame for the cotangent bundle takes the form $\{dx_\beta^1, dx_\beta^2\}$. Since $\Phi_{\alpha\beta}$ is an affine morphism, $\{\Phi_{\alpha\beta}^* \partial_{x_\beta^1}, \Phi_{\alpha\beta}^* \partial_{x_\beta^2}\}$ are affine Killing vector fields on $\mathcal{U}_\alpha$. Consequently, by Lemma 12.14, this is a real analytic frame. Thus, the dual frame for the cotangent bundle $\{\Phi_{\alpha\beta}^* dx_\beta^1, \Phi_{\alpha\beta}^* dx_\beta^2\}$ is real analytic as well. This implies that the coordinate functions $\Phi_{\alpha\beta}^* x_\beta^1$ and $\Phi_{\alpha\beta}^* x_\beta^2$ are real analytic. Consequently, $\Phi_{\alpha\beta}$ is real analytic. This shows that the Type $\mathcal{A}$ coordinate atlas is real analytic.

Suppose next that $\mathcal{M}$ has a Type $\mathcal{B}$ model. If $\dim(\mathfrak{K}(\mathcal{M})) = 2$, we may apply Theorem 12.51 to see that $\mathcal{M}$ is real analytic. We will show subsequently in Theorem 14.17 that if $\dim(\mathfrak{K}(\mathcal{M})) = 4$, then $\mathcal{M}$ is also a Type $\mathcal{A}$ geometry and the analysis performed above pertains. The only remaining possibility is that $\dim(\mathfrak{K}(\mathcal{M})) = 3$. The Riemannian and Lorentzian hyperbolic planes are real analytic. If $\mathcal{M}$ is not modeled on the Riemannian or Lorentzian hyperbolic plane, then Lemma 14.15 shows that

$$\mathfrak{K}(\mathcal{M}) = \text{span}\{e_1 := x^1\partial_{x^1} + x^2\partial_{x^2},\ e_2 := \partial_{x^2},\ e_3 := 2x^1x^2\partial_{x^1} + (x^2)^2\partial_{x^2}\}.$$

Thus the elements of $\mathfrak{K}(\mathcal{M})$ are real analytic. Note that $\{e_1, e_2\}$ is a frame for the tangent bundle. Let $e^1 := (x^1)^{-1}dx^1$ and $e^2 := (x^1)^{-1}x^2dx^1 - dx^2$ be the corresponding dual frame. Since $\Phi^*(e_1)$ and $\Phi^*(e_2)$ are real analytic, we conclude $\Phi^*(e^1)$ and $\Phi^*(e^2)$ are real analytic. We have $\Phi^*(e^1)\{\Phi^*(e_3)\} = 2\Phi^*(x^2)$ so $\Phi^*(x^2)$ is real analytic. Furthermore $\Phi^*(e^1) = d\log(\Phi^*(x^1))$, and thus $\Phi^*(x^1)$ is real analytic as well. This implies Φ is real analytic as desired. □

12.5.3 $\mathfrak{K}_\mathcal{A}$ AND $\mathfrak{K}_\mathcal{B}$-MODULES. Recall that $\mathfrak{K}_\mathcal{A} = \text{span}\{\partial_{x^1}, \partial_{x^2}\}$ is the Lie algebra of $\mathbb{R}^2$ and that $\mathfrak{K}_\mathcal{B} = \text{span}\{x^1\partial_{x^1} + x^2\partial_{x^2}, \partial_{x^2}\}$ is the Lie algebra of the $ax + b$ group. We define the associated polynomial algebras of functions by setting

$$\mathfrak{P}^\mathcal{A} := \mathbb{C}[e^{\alpha_1 x^1 + \alpha_2 x^2}, x^1, x^2]_{\alpha_i \in \mathbb{C}} \quad \text{and} \quad \mathfrak{P}^\mathcal{B} := \mathbb{C}[(x^1)^\alpha, \log(x^1), x^2]_{\alpha \in \mathbb{C}}.$$

Lemma 12.53 *Let $\mathcal{E}$ be a k-dimensional space of smooth complex functions on $\mathbb{R}^2$ (resp. $\mathbb{R}^+ \times \mathbb{R}$). If $\mathcal{E}$ is invariant under the action of $\mathfrak{K}_\mathcal{A}$ (resp. $\mathfrak{K}_\mathcal{B}$), then $\mathcal{E} \subset \mathfrak{P}^\mathcal{A}$ (resp. $\mathcal{E} \subset \mathfrak{P}^\mathcal{B}$).*

Proof. Suppose $\mathcal{E}$ is invariant under the action of $\mathfrak{K}_\mathcal{A}$. Decompose $\mathcal{E} = \oplus_{\alpha_1,\alpha_2} E_{\alpha_1,\alpha_2}$ where

$$E_{\alpha_1,\alpha_2} := \{f \in \mathcal{E} : (\partial_{x^1} - \alpha_1)^k f = 0 \text{ and } (\partial_{x^2} - \alpha_2)^k f = 0\}$$

are the simultaneous generalized eigenspaces of ∂_{x^1} and ∂_{x^2}. Let

$$f(x^1, x^2) = e^{\alpha_1 x^1 + \alpha_2 x^2} \tilde{f}(x^1, x^2) \in E_{\alpha_1,\alpha_2}.$$

We have $0 = (\partial_{x^1} - \alpha_1)^k f = e^{\alpha_1 x^1 + \alpha_2 x^2}\partial_{x^1}^k \tilde{f}$ and $0 = (\partial_{x^2} - \alpha_2)^k f = e^{\alpha_1 x^1 + \alpha_2 x^2}\partial_{x^2}^k \tilde{f}$. Consequently, $\partial_{x^1}^k \tilde{f} = 0$ and $\partial_{x^2}^k \tilde{f} = 0$. This implies $\tilde{f}$ is polynomial.

Suppose $\mathcal{E}$ is invariant under the action of $\mathfrak{K}_\mathcal{B}$. We may decompose $\mathcal{E} = \oplus_\alpha E_\alpha$ into the generalized eigenspaces of $X := x^1 \partial_{x^1} + x^2 \partial_{x^2}$ where

$$E_\alpha := \{f \in \mathcal{E} : (X - \alpha)^k f = 0\}.$$

If $(X - \alpha)^k f = 0$, then $f = (x^1)^\alpha \sum_{i \le k} \log(x^1)^i f_i(x^2)$. Because $[X, \partial_{x^2}] = -\partial_{x^2}$,

$$(X - \alpha - 1)^k \partial_{x^2} f = \partial_{x^2} (X - \alpha)^k f = 0 \quad \text{so} \quad \partial_{x^2} : E_\alpha \to E_{\alpha-1}.$$

Since $\dim(\mathcal{E}) = k$, $\partial_{x^2}^k f = 0$. Consequently, each of the $f_i(x^2)$ is polynomial in x^2. □

We complexify and set $E_\mathbb{C}(\mu, \mathcal{M}) := E(\mu, \mathcal{M}) \otimes_\mathbb{R} \mathbb{C}$ and $\mathfrak{K}_\mathbb{C}(\mathcal{M}) := \mathfrak{K}(\mathcal{M}) \otimes_\mathbb{R} \mathbb{C}$. We can take the real and imaginary parts to obtain corresponding real bases. It follows from Lemma 12.14 that the elements of $\mathfrak{K}(\mathcal{M})$ and $E(\mu, \mathcal{M})$ are real analytic if $\mathcal{M}$ determines either a Type $\mathcal{A}$ or a Type $\mathcal{B}$ geometry. We use Lemma 12.14 and Lemma 12.53 to establish the following result, which makes this observation much more specific; it will inform our discussion of $E(\mu, \mathcal{M})$ and $\mathfrak{K}(\mathcal{M})$ subsequently in Chapter 13 and Chapter 14.

Lemma 12.54   *Let $f \in E_\mathbb{C}(\mu, \mathcal{M})$. Let $\xi \in \mathfrak{K}_\mathbb{C}(\mathcal{M})$. Decompose $\xi = \xi_1 \partial_{x^1} + \xi_2 \partial_{x^2}$. If $\mathcal{M}$ is a Type $\mathcal{A}$ (resp. Type $\mathcal{B}$) model, then $\{f, \xi_1, \xi_2\} \subset \mathfrak{P}^\mathcal{A}$ (resp. $\{f, \xi_1, \xi_2\} \subset \mathfrak{P}^\mathcal{B}$).*

Proof. By Lemma 12.14, $\mathfrak{K}(\mathcal{M})$ and $E(\mu, \mathcal{M})$ are finite-dimensional $\mathfrak{K}(\mathcal{M})$-modules. If $\mathcal{M}$ is a Type $\mathcal{A}$ model, then $\mathfrak{K}_\mathcal{A}$ is a Lie subalgebra of $\mathfrak{K}(\mathcal{M})$; if $\mathcal{M}$ is a Type $\mathcal{B}$ model, then $\mathfrak{K}_\mathcal{B}$ is a Lie subalgebra of $\mathfrak{K}(\mathcal{M})$. Let $f \in E_\mathbb{C}(\mu, \mathcal{M})$. We apply Lemma 12.53 to see $f \in \mathfrak{P}^\mathcal{A}$ if $\mathcal{M}$ is a Type $\mathcal{A}$ model and $f \in \mathfrak{P}^\mathcal{B}$ if $\mathcal{M}$ is a Type $\mathcal{B}$ model. Suppose that

$$\xi = \xi_1 \partial_{x^1} + \xi_2 \partial_{x^2} \in \mathfrak{K}_\mathbb{C}(\mathcal{M})$$

and that $\mathcal{M}$ is a Type $\mathcal{A}$ model. Then $[\partial_{x^i}, \xi] = \partial_{x^i} \xi_1 \partial_{x^1} + \partial_{x^i} \xi_2 \partial_{x^2}$ and, consequently, ξ_i belongs to $\mathfrak{P}^\mathcal{A}$. Let $X = x^1 \partial_{x^1} + x^2 \partial_{x^2}$. If $\mathcal{M}$ is a Type $\mathcal{B}$ model, then

$$[X, \xi] = (X(\xi_1) - \xi_1) \partial_{x^1} + (X(\xi_2) - \xi_2) \partial_{x^2} \quad \text{and} \quad [\partial_{x^2}, \xi] = \partial_{x^2}(\xi_1) \partial_{x^1} + \partial_{x^2}(\xi_2) \partial_{x^2}.$$

Thus although the module action of $\mathfrak{K}_\mathcal{B}$ on $\mathfrak{K}(\mathcal{M})$ is a bit different than the action on C^∞, the analysis of Lemma 12.53 again yields $\xi_i \in \mathfrak{P}^\mathcal{B}$. □

CHAPTER 13

The Geometry of Type $\mathcal{A}$ Models

In Chapter 13, we shall report on work of Brozos-Vázquez et al. [9–12] and Gilkey and Valle-Regueiro [35] that deals with the affine quasi-Einstein equation, work of Brozos-Vázquez, García-Río, and Gilkey [7] that deals with affine Killing vector fields and gradient Ricci solitons, work of Brozos-Vázquez, García-Río, and Gilkey [6] that deals with moduli spaces, and work of other authors as cited. We shall assume that $\mathcal{M} = (\mathbb{R}^2, \nabla)$ is a Type $\mathcal{A}$ model unless otherwise noted; this means that the Christoffel symbols of ∇ are constant. We begin by recalling some results from Chapter 12 that will play a crucial role in what follows.

Observation 13.1   We have defined $\mathcal{Q}(\mathcal{M}) = \{f \in C^\infty(\mathbb{R}^2) : \mathcal{H}f + f\rho_s = 0\}$ for a surface $\mathcal{M}$. Let g be a smooth function which defines a strong projective equivalence between two affine manifolds $\mathcal{M}$ and $^g\mathcal{M}$, i.e.,

$$
\begin{aligned}
{}^g\Gamma_{11}{}^1 &= 2\partial_{x^1}g + \Gamma_{11}{}^1, & {}^g\Gamma_{11}{}^2 &= \Gamma_{11}{}^2, \\
{}^g\Gamma_{12}{}^1 &= \partial_{x^2}g + \Gamma_{12}{}^1, & {}^g\Gamma_{12}{}^2 &= \partial_{x^1}g + \Gamma_{12}{}^2, \\
{}^g\Gamma_{22}{}^1 &= \Gamma_{22}{}^1, & {}^g\Gamma_{22}{}^2 &= 2\partial_{x^2}g + \Gamma_{22}{}^2 .
\end{aligned}
$$

We then have by Theorem 12.29 that $\mathcal{Q}(^g\mathcal{M}) = e^g\mathcal{Q}(\mathcal{M})$. Suppose that $\mathcal{M}$ is a Type $\mathcal{A}$ model. We will show in Lemma 13.6 that $\mathcal{M}$ is strongly projectively flat. Consequently, by Theorem 12.34, $\dim(\mathcal{Q}(\mathcal{M})) = 3$. The space of functions $\mathcal{Q}(\mathcal{M})$ is a complete invariant of Type $\mathcal{A}$ models; by Theorem 12.34, if $\mathcal{M}_i$ are Type $\mathcal{A}$ models, then $\mathcal{M}_1 = \mathcal{M}_2$ if and only if $\mathcal{Q}(\mathcal{M}_1) = \mathcal{Q}(\mathcal{M}_2)$. Let Ψ be a diffeomorphism from M_1 to M_2. Then Ψ intertwines $\mathcal{M}_1$ and $\mathcal{M}_2$ if and only if $\Psi^*\mathcal{Q}(\mathcal{M}_2) = \mathcal{Q}(\mathcal{M}_1)$. By Lemma 12.53, if $\mathcal{M}$ is a Type $\mathcal{A}$ model, then $\mathcal{Q}(\mathcal{M})$ is a 3-dimensional space which is spanned by products of linear exponentials and polynomials and which is invariant under the action of ∂_{x^i}. This will inform our discussion.

In Section 13.1, we present some foundational results. In Section 13.2, see Definition 13.7, we will present some basic examples of Type $\mathcal{A}$ models. We will determine the Ricci tensor, the algebra of affine Killing vector fields, and the spaces $\mathcal{Q}$ for these examples. In Section 13.3, we will use the analytic facts outlined in Observation 13.1 to show that any Type $\mathcal{A}$ model is linearly equivalent to one of the examples discussed in Section 13.2. The α invariant of Lemma 12.22 is determined when the Ricci tensor has rank 1. Theorem 12.35 gives an ansatz for determining the geodesic structure of a strongly projectively flat surface. We will use this ansatz subsequently to determine which Type $\mathcal{A}$ models are geodesically complete. In Section 13.4, we discuss moduli spaces of Type $\mathcal{A}$ models.

13.1 TYPE $\mathcal{A}$: FOUNDATIONAL RESULTS AND BASIC EXAMPLES

In Section 13.1.1, we discuss various results concerning the Ricci tensor for Type $\mathcal{A}$ geometries. In Section 13.1.2, we give a useful criteria to ensure the Ricci tensor has rank 1 and show such a tensor is recurrent. In Section 13.1.3, we relate the rank of the space of affine Killing vector fields to the Ricci tensor (see Lemma 13.4); we will subsequently improve this result in Corollary 13.25, but this weaker form will be useful in our analysis at this stage. In Section 13.1.4, we show linear equivalence and affine equivalence are equivalent concepts if the Ricci tensor has rank 2; this fails if the Ricci tensor is degenerate (see Remark 13.13). In Section 13.1.5, we will give a direct combinatorial proof that any Type $\mathcal{A}$ geometry is linearly strongly projectively flat. This will permit us to use various results concerning the solution space to the affine quasi-Einstein equation $\mathcal{Q}$ which were established in Chapter 12.

13.1.1 THE RICCI TENSOR.

Lemma 13.2   *Let $\mathcal{M} = (\mathbb{R}^2, \nabla)$ be a Type $\mathcal{A}$ model.*

1. *ρ is symmetric and $\nabla \rho$ is totally symmetric.*
2. *Let $f \in E(\mu, \mathcal{M})$ for $\mu \neq -1$. Then $R_{12}(df) = 0$.*
3. *Let f be an affine gradient Ricci soliton. Then $R_{12}(df) = 0$.*

Proof. Let $\mathcal{M}$ be a Type $\mathcal{A}$ model. We show that ρ is symmetric by computing:

$$
\begin{aligned}
\rho_{11} &= (\Gamma_{11}{}^1 - \Gamma_{12}{}^2)\Gamma_{12}{}^2 + \Gamma_{11}{}^2(\Gamma_{22}{}^2 - \Gamma_{12}{}^1), \\
\rho_{12} &= \rho_{21} = \Gamma_{12}{}^1\Gamma_{12}{}^2 - \Gamma_{11}{}^2\Gamma_{22}{}^1, \\
\rho_{22} &= -(\Gamma_{12}{}^1)^2 + \Gamma_{22}{}^2\Gamma_{12}{}^1 + (\Gamma_{11}{}^1 - \Gamma_{12}{}^2)\Gamma_{22}{}^1 \, .
\end{aligned}
\tag{13.1.a}
$$

We show that $\nabla \rho$ is totally symmetric and establish Assertion 1 by computing:

$$
\begin{aligned}
\rho_{11;1} &= 2\{-(\Gamma_{11}{}^1)^2\Gamma_{12}{}^2 + \Gamma_{11}{}^1(\Gamma_{11}{}^2(\Gamma_{12}{}^1 - \Gamma_{22}{}^2) + (\Gamma_{12}{}^2)^2) \\
&\quad + \Gamma_{11}{}^2(\Gamma_{11}{}^2\Gamma_{22}{}^1 - \Gamma_{12}{}^1\Gamma_{12}{}^2)\}, \\
\rho_{12;1} &= \rho_{21;1} = \rho_{11;2} = 2\left(\Gamma_{11}{}^2\left((\Gamma_{12}{}^1)^2 - \Gamma_{12}{}^1\Gamma_{22}{}^2 + \Gamma_{12}{}^2\Gamma_{22}{}^1\right) - \Gamma_{11}{}^1\Gamma_{12}{}^1\Gamma_{12}{}^2\right), \\
\rho_{12;2} &= \rho_{21;2} = \rho_{22;1} = 2\left(\Gamma_{12}{}^2(-\Gamma_{11}{}^1\Gamma_{22}{}^1 - \Gamma_{12}{}^1\Gamma_{22}{}^2 + \Gamma_{12}{}^2\Gamma_{22}{}^1) + \Gamma_{11}{}^2\Gamma_{12}{}^1\Gamma_{22}{}^1\right), \\
\rho_{22;2} &= 2\{\Gamma_{22}{}^1(\Gamma_{22}{}^2(\Gamma_{12}{}^2 - \Gamma_{11}{}^1) + \Gamma_{11}{}^2\Gamma_{22}{}^1) + (\Gamma_{12}{}^1)^2\Gamma_{22}{}^2 \\
&\quad - \Gamma_{12}{}^1(\Gamma_{12}{}^2\Gamma_{22}{}^1 + (\Gamma_{22}{}^2)^2)\} \, .
\end{aligned}
$$

Let $f \in E(\mu, \mathcal{M})$. We covariantly differentiate the relation $\mathcal{H}f = \mu f \rho$ to conclude that $f_{;jki} = \mu\{f_{;i}\rho_{jk} + f\rho_{jk;i}\}$. Since $\nabla\rho$ is totally symmetric, anti-symmetrize in i and k to see

$$R_{kij}{}^{\ell} f_{;\ell} = f_{;jki} - f_{;jik} = \mu\{f_{;i}\rho_{jk} - f_{;k}\rho_{ij}\}.$$

We take $i = 2$ and $k = 1$ and unpack this expression after setting $\rho_{12} = \rho_{21}$:

$$(j = 1) \quad R_{121}{}^{1} f_{;1} + R_{121}{}^{2} f_{;2} = \mu(f_{;2}\rho_{11} - f_{;1}\rho_{12})$$
$$= \mu\{f_{;2}\rho_{11} - f_{;1}\rho_{21}\} = \mu(-R_{121}{}^{1} f_{;1} + R_{211}{}^{2} f_{;2}),$$
$$(j = 2) \quad R_{122}{}^{1} f_{;1} + R_{122}{}^{2} f_{;2} = \mu(f_{;2}\rho_{12} - f_{;1}\rho_{22})$$
$$= \mu(f_{;2}\rho_{21} - f_{;1}\rho_{22}) = \mu(-R_{122}{}^{1} f_{;1} + R_{212}{}^{2} f_{;2}).$$

This shows $(\mu + 1)R_{12j}{}^{\ell} f_{;\ell} = 0$ for $j = 1, 2$ or, equivalently, $R_{12}(df) = 0$ since $\mu \neq -1$. This establishes Assertion 2; the proof of Assertion 3 follows by applying the same argument to the identity $f_{;jki} = -\rho_{jk;i}$. □

13.1.2 RECURRENT TENSORS. The following is a useful observation.

Lemma 13.3 *Let $\mathcal{M}$ be a Type $\mathcal{A}$ model which is not flat.*

1. *The following conditions are equivalent:*
   (a) $\rho = \rho_{22}dx^2 \otimes dx^2$.
   (b) $\Gamma_{11}{}^2 = 0$ and $\Gamma_{12}{}^2 = 0$.
   (c) $\rho = \{\Gamma_{12}{}^1(\Gamma_{22}{}^2 - \Gamma_{12}{}^1) + \Gamma_{11}{}^1\Gamma_{22}{}^1\}dx^2 \otimes dx^2$.
2. *$\mathcal{M}$ is recurrent if and only if $\mathrm{Rank}(\rho) = 1$.*

Proof. We use Equation (13.1.a) to compute ρ. We first show that Assertion 1-a implies Assertion 1-b. We distinguish cases.

Case 1. Suppose that $\Gamma_{22}{}^1$ is non-zero. By rescaling, we may suppose $\Gamma_{22}{}^1 = 1$. To ensure $\rho_{12} = 0$, we set $\Gamma_{11}{}^2 = \Gamma_{12}{}^1\Gamma_{12}{}^2$ and obtain $\rho_{11} = \Gamma_{12}{}^2\rho_{22}$. Since $\rho_{22} \neq 0$, $\Gamma_{12}{}^2 = 0$, and hence $\Gamma_{11}{}^2 = 0$ as well as desired.

Case 2. Suppose that $\Gamma_{22}{}^1 = 0$. Then $0 = \rho_{12} = \Gamma_{12}{}^1\Gamma_{12}{}^2$. Since $\rho_{22} = \Gamma_{12}{}^1(\Gamma_{22}{}^2 - \Gamma_{12}{}^1)$, $\Gamma_{12}{}^1 \neq 0$. Thus $\Gamma_{12}{}^2 = 0$, $\rho_{11} = \Gamma_{11}{}^2(\Gamma_{22}{}^2 - \Gamma_{12}{}^1)$, and $\rho_{22} = \Gamma_{12}{}^1(\Gamma_{22}{}^2 - \Gamma_{12}{}^1)$. Consequently, $\Gamma_{11}{}^2 = 0$ as desired.

We have shown that Assertion 1-a implies Assertion 1-b. We make a direct computation to show that Assertion 1-b implies Assertion 1-c. It is immediate that Assertion 1-c implies Assertion 1-a. This completes the proof of Assertion 1. We now establish Assertion 2. If $\mathrm{Rank}(\rho) = 1$, we can change coordinates to assume $\rho = \rho_{22}dx^2 \otimes dx^2$. We then use Assertion 1 to see that $\Gamma_{11}{}^2 = \Gamma_{12}{}^2 = 0$. This yields

$$\rho = \rho_{22}dx^2 \otimes dx^2 \quad \text{and} \quad \nabla\rho = -2\Gamma_{22}{}^2\rho_{22}dx^2 \otimes dx^2 \otimes dx^2.$$

Consequently, ρ is recurrent. Conversely, suppose ρ is recurrent. Make a linear change of co-ordinates to diagonalize ρ. Since $\mathcal{M}$ is not flat, ρ does not vanish identically and thus we may assume the notation is chosen so $\rho_{22} \neq 0$. Again we distinguish cases setting $\rho_{12} = 0$.

Case 1. Suppose that $\Gamma_{22}{}^1 \neq 0$. We rescale to assume $\Gamma_{22}{}^1 = 1$ and set $\Gamma_{11}{}^2 = \Gamma_{12}{}^1\Gamma_{12}{}^2$ to ensure $\rho_{12} = 0$. We obtain $\rho_{11} = \Gamma_{12}{}^2\rho_{22}$ and $\rho_{12;2} = -2\Gamma_{12}{}^2\rho_{22}$. Since $\rho_{22} \neq 0$ and since $\rho_{12;2} = \omega_2\rho_{12} = 0$, we obtain $\Gamma_{12}{}^2 = 0$. Thus $\rho_{11} = 0$ and $\mathrm{Rank}(\rho) = 1$ as desired.

Case 2. Suppose that $\Gamma_{22}{}^1 = 0$. Setting $\rho_{12} = 0$ and $\rho_{22} \neq 0$ then yields $\Gamma_{12}{}^2 = 0$. We have $0 = \omega_1\rho_{12} = \rho_{12;1} = -2\Gamma_{11}{}^2\rho_{22}$ and thus $\Gamma_{11}{}^2 = 0$. We apply Assertion 1 to see $\mathrm{Rank}(\rho) = 1$ as desired. □

13.1.3 AFFINE KILLING VECTOR FIELDS.
For Type $\mathcal{A}$ models, there is a close link between the rank of the Ricci tensor and the dimension of the space of affine Killing vector fields. This is not the case for the Type $\mathcal{B}$ models.

Lemma 13.4   *Let $\mathcal{M}$ be a Type $\mathcal{A}$ model.*

1. *If $\mathrm{Rank}(\rho) = 0$, then $\dim(\mathfrak{K}(\mathcal{M})) = 6$.*
2. *If $\mathrm{Rank}(\rho) = 1$, then $\dim(\mathfrak{K}(\mathcal{M})) \leq 4$.*
3. *If $\mathrm{Rank}(\rho) = 2$, then $\dim(\mathfrak{K}(\mathcal{M})) = 2$.*

We shall show in Corollary 13.25 that equality holds in Assertion 2 of Lemma 13.4; $\dim(\mathfrak{K}(\mathcal{M})) = 4$ if and only if the Ricci tensor of $\mathcal{M}$ has rank 1.

Proof. By Lemma 12.14, $\dim(\mathfrak{K}(\mathcal{M})) \leq 6$. Let $\mathbb{A} = (\mathbb{R}^2, \Gamma_0)$ be the affine plane where all the Christoffel symbols vanish identically. Let (a_i^j) in $M_2(\mathbb{R})$ be a constant matrix and let (b^1, b^2) in $\mathbb{R}^2$. One verifies $a_i^j x^i \partial_{x^j} + b^k \partial_{x^k} \in \mathfrak{K}(\mathbb{A})$ and thus $\dim(\mathfrak{K}(\mathbb{A})) = 6$. Let $\mathcal{M}$ be a Type $\mathcal{A}$ model with $\mathrm{Rank}(\rho) = 0$. Then $\rho = 0$ so by Lemma 12.1, $\mathcal{M}$ is flat. Thus $\dim(\mathfrak{K}_P(\mathcal{M})) = 6$ for every point $P \in \mathbb{R}^2$. Since $\mathbb{R}^2$ is simply connected, $\dim(\mathfrak{K}(\mathcal{M})) = 6$ which establishes Assertion 1.

Suppose $\mathrm{Rank}(\rho) = 1$. By Lemma 13.3, we can make a linear change of coordinates to assume ρ is a non-zero multiple of $dx^2 \otimes dx^2$. Let $X = a^1(x^1, x^2)\partial_{x^1} + a^2(x^1, x^2)\partial_{x^2}$ be an affine Killing vector field. By subtracting appropriate multiples of ∂_{x^1} and ∂_{x^2}, we may assume $X(0) = 0$. Then

$$
\begin{aligned}
0 &= (\mathcal{L}_X\rho)(\partial_{x^1}, \partial_{x^2}) = X(\rho(\partial_{x^1}, \partial_{x^2})) - \rho([X, \partial_{x^1}], \partial_{x^2}) - \rho(\partial_{x^1}, [X, \partial_{x^2}]) \\
&= 0 + (\partial_{x^1}a^2)\rho(\partial_{x^2}, \partial_{x^2}) + 0 \Rightarrow \partial_{x^1}a^2 = 0, \\
0 &= (\mathcal{L}_X\rho)(\partial_{x^2}, \partial_{x^2}) = X(\rho(\partial_{x^2}, \partial_{x^2})) - 2\rho([X, \partial_{x^2}], \partial_{x^2}) \\
&= 0 + 2(\partial_{x^2}a^2)\rho(\partial_{x^2}, \partial_{x^2}) \Rightarrow \partial_{x^2}a^2 = 0.
\end{aligned}
$$

Thus a^2 is constant. Since $X(0) = 0$, there is no ∂_{x^2} dependence and $X = a^1(x^1, x^2)\partial_{x^1}$. Since $X(0) = 0$ and X is determined by $X(0)$ and $dX(0)$, the set of such X forms at most a 2-dimensional space so $\dim(\mathfrak{K}(\mathcal{M})) \leq 4$. This proves Assertion 2.

Suppose that ρ has rank 2 so ρ is non-degenerate. Let X be an affine Killing vector field with $X(0) = 0$. We wish to show X vanishes identically and thus $\mathfrak{K}(\mathcal{M}) = \text{span}\{\partial_{x^1}, \partial_{x^2}\}$. Suppose to the contrary that there exists an affine Killing vector field which is non-trivial with $X(0) = 0$. Let Φ_t^X be the (local) flow of X. The relations $(\Phi_t^X)^* \rho = \rho$ and $\Phi_t^X(0) = \text{Id}$ imply that $\Phi_t^X \in \text{SO}(\rho)$. If ρ is definite, we can choose coordinates to assume that

$$\rho = \pm\{(dx^1)^2 + (dx^2)^2\} \quad \text{and} \quad \text{SO}(\rho) = \text{SO}(2).$$

Consequently, after rescaling X if necessary, we may identify Φ_t^X with a rotation through an angle t. No non-zero constant Christoffel symbol Γ is preserved by the action of $\text{SO}(2)$ (two indices are down and one is up), so this provides the desired contradiction. The argument is analogous if ρ is Lorentzian if we replace $\text{SO}(2)$ by $\text{SO}(1, 1)$. Assertion 3 follows. $\qquad \square$

13.1.4 LINEAR EQUIVALENCE. In the context of Type $\mathcal{A}$ models with non-degenerate Ricci tensor, linear equivalence and affine equivalence are the same concept. This vastly simplifies the analysis.

Theorem 13.5   *Let $\mathcal{M}_i$ be Type $\mathcal{A}$ models such that $\rho_{\mathcal{M}_i}$ are non-degenerate. Then $\mathcal{M}_1$ is linearly equivalent to $\mathcal{M}_2$ if and only if $\mathcal{M}_1$ is affinely equivalent to $\mathcal{M}_2$.*

Proof. Although this follows from work of Brozos-Vázquez, García-Río, and Gilkey [7], we give a slightly different derivation to keep our present treatment as self-contained as possible. Let $\mathcal{M}_i$ be Type $\mathcal{A}$ models with non-degenerate Ricci tensors. By Lemma 13.4, $\dim(\mathfrak{K}(\mathcal{M}_i)) = 2$. If ϕ is an affine transformation from $\mathcal{M}_1$ to $\mathcal{M}_2$, then $\phi_* \mathfrak{K}(\mathcal{M}_1) = \mathfrak{K}(\mathcal{M}_2)$. Thus $\phi_*(\partial_{x^i}) = a_i^j \partial_{x^j}$ for some $(a_i^j) \in \text{GL}(2, \mathbb{R})$. By subtracting an appropriate translation, we may assume $\phi(0) = 0$. It now follows that $\phi \in \text{GL}(2, \mathbb{R})$. $\qquad \square$

In Remark 13.13, we will show that Theorem 13.5 fails if the Ricci tensor is not assumed to be non-degenerate.

13.1.5 LINEARLY STRONGLY PROJECTIVELY FLAT. We say that a Type $\mathcal{A}$ model $\mathcal{M}$ is *linearly strongly projectively flat* if there exists a linear function g which provides projective equivalence from $\mathcal{M}$ to a flat geometry $\mathcal{M}_f$; note that $\mathcal{M}_f$ is again a Type $\mathcal{A}$ model in this setting.

Lemma 13.6   *Let $\mathcal{M} = (\mathbb{R}^2, \nabla)$ be a Type $\mathcal{A}$ model. There exists a linear function of the form $g(x^1, x^2) = a_1 x^1 + a_2 x^2$ which provides a strong projective equivalence from $\mathcal{M}$ to a flat Type $\mathcal{A}$ model and which satisfies $e^{-g} \in \mathcal{Q}(\mathcal{M})$. Thus, in particular, $\mathcal{M}$ is linearly strongly projectively flat.*

Proof. Although Lemma 13.2 and Theorem 12.11 imply that $\mathcal{M}$ is strongly projectively flat, we do not obtain that this can be done by choosing $\omega = a_1 dx^1 + a_2 dx^2$ to be constant. We follow the discussion in Gilkey and Valle-Regueiro [35]. We work modulo linear equivalence.

We use Equation (13.1.a) to study the Ricci tensor ρ of $\mathcal{M}$. Let $g(x^1, x^2) = a_1 x^1 + a_2 x^2$ for $(a_1, a_2) \in \mathbb{R}^2$.

Case 1. Suppose $\Gamma_{11}{}^2 \neq 0$. Rescale x^2 to ensure $\Gamma_{11}{}^2 = 1$. We have

$$(\rho_{g\,\mathcal{M}})_{11} = a_1^2 + a_1 \Gamma_{11}{}^1 - \Gamma_{12}{}^1 + \Gamma_{11}{}^1 \Gamma_{12}{}^2 - (\Gamma_{12}{}^2)^2 + a_2 + \Gamma_{22}{}^2 \,.$$

We set $a_2 := -a_1^2 - a_1 \Gamma_{11}{}^1 + \Gamma_{12}{}^1 - \Gamma_{11}{}^1 \Gamma_{12}{}^2 + (\Gamma_{12}{}^2)^2 - \Gamma_{22}{}^2$ to ensure $(\rho_{g\,\mathcal{M}})_{11} = 0$. Then

$$
\begin{aligned}
(\rho_{g\,\mathcal{M}})_{12} &= -\Gamma_{22}{}^1 - (a_1 + \Gamma_{12}{}^2)(a_1^2 + a_1 \Gamma_{11}{}^1 - 2\Gamma_{12}{}^1 + \Gamma_{11}{}^1 \Gamma_{12}{}^2 - (\Gamma_{12}{}^2)^2 + \Gamma_{22}{}^2), \\
(\rho_{g\,\mathcal{M}})_{22} &= (-a_1 - \Gamma_{11}{}^1 + \Gamma_{12}{}^2)(\rho_{g\,\mathcal{M}})_{12} \,.
\end{aligned}
$$

Since $(\rho_{g\,\mathcal{M}})_{12}$ is cubic in a_1 with non-zero leading coefficient, we can find a_1 so $(\rho_{g\,\mathcal{M}})_{12} = 0$. Since $(\rho_{g\,\mathcal{M}})_{12}$ divides $(\rho_{g\,\mathcal{M}})_{22}$, $(\rho_{g\,\mathcal{M}})_{22} = 0$ as well so $\rho_{g\,\mathcal{M}} = 0$.

Case 2. Suppose $\Gamma_{11}{}^2 = 0$. Since the argument is the same as that given in Case 1 if $\Gamma_{22}{}^1 \neq 0$, we may assume $\Gamma_{22}{}^1 = 0$ as well. We make a direct computation to see that taking $a_1 = -\Gamma_{12}{}^2$ and $a_2 = -\Gamma_{12}{}^1$ yields $\rho_{g\,\mathcal{M}} = 0$.

We have chosen a linear function g so that $\rho_{g\,\mathcal{M}} = 0$. This implies ${}^g\mathcal{M}$ is flat. Consequently, $\mathcal{M} = {}^{-g}\{{}^g\mathcal{M}\}$ is linearly strongly projectively flat. Since $\mathcal{Q}(\mathcal{M}) = e^{-g} \mathcal{Q}({}^g\mathcal{M})$ and since $\mathbb{1} \in \mathcal{Q}({}^g\mathcal{M})$, we conclude $e^{-g} \in \mathcal{Q}(\mathcal{M})$. $\square$

13.2   TYPE $\mathcal{A}$: DISTINGUISHED GEOMETRIES

In Section 13.2.1 (see Definition 13.7), we define 15 families of Type $\mathcal{A}$ models that will form the focus of our investigations in Chapter 13. We will show subsequently in Section 13.3 (see Theorem 13.22) that any Type $\mathcal{A}$ model is linearly equivalent to one of these examples, and thus establishing their basic properties is crucial. Section 13.2.2 gives the Ricci tensor, Section 13.2.3 gives the solution space $\mathcal{Q}$ of the affine quasi-Einstein equation for $\mu = -1$, and Section 13.2.4 gives the spaces $E(\mu)$ for $\mu \neq -1$. Some useful local affine embeddings, immersions, and equivalences are given in Section 13.2.5. Section 13.2.6 treats affine gradient Ricci solitons. The space of affine Killing vector fields is given in Section 13.2.7. Section 13.2.8 treats invariant and parallel tensors of Type (1,1) for these examples. We thought it best to gather all the basic information concerning these examples in one section for the convenience of the reader; we will use these computations throughout the rest of this section.

13.2.1 DEFINING TYPE $\mathcal{A}$ MODELS. We introduce the following notational conventions.

Definition 13.7 For $\vec{\xi} := (\xi_1, \xi_2, \xi_3, \xi_4, \xi_5, \xi_6) \in \mathbb{R}^6$, let

$$
\begin{aligned}
\Gamma(\vec{\xi}) \quad &:= \quad \Gamma(\xi_1, \xi_2, \xi_3, \xi_4, \xi_5, \xi_6) \\
&= \quad \{\Gamma_{11}{}^1 = \xi_1, \ \Gamma_{11}{}^2 = \xi_2, \ \Gamma_{12}{}^1 = \xi_3, \\
&\qquad\quad \Gamma_{12}{}^2 = \xi_4, \ \Gamma_{22}{}^1 = \xi_5, \ \Gamma_{22}{}^2 = \xi_6\}, \\
\mathcal{M}(\vec{\xi}) \quad &:= \quad \mathcal{M}(\xi_1, \xi_2, \xi_3, \xi_4, \xi_5, \xi_6) = (\mathbb{R}^2, \Gamma(\vec{\xi})).
\end{aligned}
\tag{13.2.a}
$$

Define the following Type $\mathcal{A}$ affine models.

$\mathcal{M}_i^6$: $\quad \mathcal{M}_0^6 := \mathcal{M}(0,0,0,0,0,0), \qquad \mathcal{M}_1^6 := \mathcal{M}(1,0,0,1,0,0),$

$\qquad \mathcal{M}_2^6 := \mathcal{M}(-1,0,0,0,0,1), \qquad \mathcal{M}_3^6 := \mathcal{M}(0,0,0,0,0,1),$

$\qquad \mathcal{M}_4^6 := \mathcal{M}(0,0,0,0,1,0), \qquad \mathcal{M}_5^6 := \mathcal{M}(1,0,0,1,-1,0).$

$\mathcal{M}_i^4$: $\quad \mathcal{M}_1^4 := \mathcal{M}(-1,0,1,0,0,2), \qquad \mathcal{M}_2^4(c_1) := \mathcal{M}(-1,0,c_1,0,0,1+2c_1),$

$\qquad \mathcal{M}_3^4(c_1) := \mathcal{M}(0,0,c_1,0,0,1+2c_1), \quad \mathcal{M}_4^4(c) := \mathcal{M}(0,0,1,0,c,2),$

$\qquad \mathcal{M}_5^4(c) := \mathcal{M}(1,0,0,0,1+c^2,2c),$

$\qquad$ where $c \in \mathbb{R}$ and $c_1 \notin \{0,-1\}$.

$\mathcal{M}_i^2$: $\quad \mathcal{M}_1^2(a_1, a_2) := \mathcal{M}\left(\frac{a_1^2 + a_2 - 1, a_1^2 - a_1, a_1 a_2, a_1 a_2, a_2^2 - a_2, a_1 + a_2^2 - 1}{a_1 + a_2 - 1} \right),$

$\qquad \mathcal{M}_2^2(b_1, b_2) := \mathcal{M}\left(1 + b_1, 0, b_2, 1, \frac{1+b_2^2}{b_1 - 1}, 0 \right),$

$\qquad \mathcal{M}_3^2(c_2) := \mathcal{M}(2,0,0,1,c_2,1), \qquad \mathcal{M}_4^2(\pm 1) := \mathcal{M}(2,0,0,1,\pm 1,0),$

$\qquad$ where $a_1 a_2 \neq 0$, $a_1 + a_2 \neq 1$, $b_1 \neq 1$, $(b_1, b_2) \neq (0,0)$, and $c_2 \neq 0$.

The notation is chosen (see Lemma 13.8 and the computations of Section 13.2.7) so that

$$
\dim(\mathfrak{K}(\mathcal{M}_i^\nu(\cdot))) = \nu \quad \text{and} \quad \mathrm{Rank}(\rho_{\mathcal{M}_i^\nu(\cdot)}) = \left\{ \begin{array}{ll} 0 & \text{if } \nu = 6 \\ 1 & \text{if } \nu = 4 \\ 2 & \text{if } \nu = 2 \end{array} \right\}.
$$

13.2.2 THE RICCI TENSOR.

Lemma 13.8 *Adopt the notation of Definition 13.7.*

1. *The Ricci tensor of the geometries $\mathcal{M}_i^6$ vanishes. These geometries are flat.*

2. *The Ricci tensor of the geometries $\mathcal{M}_i^4$ has rank 1.*

$$
\begin{aligned}
\rho_{\mathcal{M}_1^4} &= dx^2 \otimes dx^2, & \rho_{\mathcal{M}_2^4(c_1)} &= (c_1 + c_1^2)dx^2 \otimes dx^2, \\
\rho_{\mathcal{M}_3^4(c_1)} &= (c_1 + c_1^2)dx^2 \otimes dx^2, & \rho_{\mathcal{M}_4^4(c)} &= dx^2 \otimes dx^2, \\
\rho_{\mathcal{M}_5^4(c)} &= (1 + c^2)dx^2 \otimes dx^2.
\end{aligned}
$$

These geometries are recurrent. Let $\Theta := dx^2 \otimes dx^2 \otimes dx^2$.

$$\nabla \rho_{\mathcal{M}_1^4} = -4\Theta, \qquad\qquad \nabla \rho_{\mathcal{M}_2^4(c_1)} = -2(1 + 2c_1)(c_1 + c_1^2)\Theta,$$
$$\nabla \rho_{\mathcal{M}_3^4(c_1)} = -2(1 + 2c_1)(c_1 + c_1^2)\Theta, \quad \nabla \rho_{\mathcal{M}_4^4(c)} = -4\Theta,$$
$$\nabla \rho_{\mathcal{M}_5^4(c)} = -4c(1 + c^2)\Theta.$$

Since $c_1 + c_1^2 \neq 0$, *only* $\mathcal{M}_2^4(-\frac{1}{2})$, $\mathcal{M}_3^4(-\frac{1}{2})$, *and* $\mathcal{M}_5^4(0)$ *are affine symmetric spaces.*

3. *The Ricci tensor of the geometries* $\mathcal{M}_i^2$ *has rank* 2.

$$\rho_{\mathcal{M}_1^2(a_1,a_2)} = \begin{pmatrix} \frac{(a_1-1)a_1}{a_1+a_2-1} & \frac{a_1 a_2}{a_1+a_2-1} \\ \frac{a_1 a_2}{a_1+a_2-1} & \frac{(a_2-1)a_2}{a_1+a_2-1} \end{pmatrix}, \quad \det(\rho_{\mathcal{M}_1^2(a_1,a_2)}) = -\frac{a_1 a_2}{a_1+a_2-1},$$

$$\rho_{\mathcal{M}_2^2(b_1,b_2)} = \begin{pmatrix} b_1 & b_2 \\ b_2 & \frac{b_2^2+b_1}{b_1-1} \end{pmatrix}, \qquad \det(\rho_{\mathcal{M}_2^2(b_1,b_2)}) = \frac{b_1^2+b_2^2}{b_1-1},$$

$$\rho_{\mathcal{M}_3^2(c_2)} = dx^1 \otimes dx^1 + c_2 dx^2 \otimes dx^2, \quad \rho_{\mathcal{M}_4^2}(\pm 1) = dx^1 \otimes dx^1 \pm dx^2 \otimes dx^2.$$

$$(\rho_{\mathcal{M}_1^2(a_1,a_2)})_{11;1} = -\frac{2a_1(a_1^2-1)}{a_1+a_2-1}, \qquad\qquad (\rho_{\mathcal{M}_1^2(a_1,a_2)})_{11;2} = -\frac{2a_1^2 a_2}{a_1+a_2-1},$$
$$(\rho_{\mathcal{M}_1^2(a_1,a_2)})_{12;2} = -\frac{2a_1 a_2^2}{a_1+a_2-1}, \qquad\qquad (\rho_{\mathcal{M}_1^2(a_1,a_2)})_{22;2} = -\frac{2a_2(a_2^2-1)}{a_1+a_2-1},$$
$$(\rho_{\mathcal{M}_2^2(b_1,b_2)})_{11;1} = -2b_1(b_1 + 1), \qquad\qquad (\rho_{\mathcal{M}_2^2(b_1,b_2)})_{11;2} = -2(b_1 + 1)b_2,$$
$$(\rho_{\mathcal{M}_2^2(b_1,b_2)})_{12;2} = -\frac{2b_1(b_2^2+1)}{b_1-1}, \qquad\qquad (\rho_{\mathcal{M}_2^2(b_1,b_2)})_{22;2} = -\frac{2b_2(b_2^2+1)}{b_1-1},$$
$$(\rho_{\mathcal{M}_3^2(c_2)})_{11;1} = -4, \qquad\qquad\qquad (\rho_{\mathcal{M}_3^2(c_2)})_{11;2} = 0,$$
$$(\rho_{\mathcal{M}_3^2(c_2)})_{12;2} = -2c_2, \qquad\qquad\qquad (\rho_{\mathcal{M}_3^2(c_2)})_{22;2} = -2c_2,$$
$$(\rho_{\mathcal{M}_4^2}(\pm 1))_{11;1} = -4, \qquad\qquad\qquad (\rho_{\mathcal{M}_4^2}(\pm 1))_{11;2} = 0,$$
$$(\rho_{\mathcal{M}_4^2}(\pm 1))_{12;2} = \mp 2, \qquad\qquad\qquad (\rho_{\mathcal{M}_4^2}(\pm 1))_{22;2} = 0.$$

None of the geometries where $\mathrm{Rank}(\rho) = 2$ *are affine symmetric spaces.*

Proof. The desired result follows by a direct computation; since $c_1 \notin \{0, -1\}$, $a_1 a_2 \neq 0$, and $c_2 \neq 0$, ρ has the correct rank.

13.2.3 THE SOLUTION SPACE $\mathcal{Q}$.

Lemma 13.9 *Adopt the notation of Definition* 13.7.

$\mathcal{Q}(\mathcal{M}_i^6)$: $\mathcal{Q}(\mathcal{M}_0^6) = \mathrm{span}\{1, x^1, x^2\}, \qquad \mathcal{Q}(\mathcal{M}_1^6) = \mathrm{span}\{1, e^{x^1}, x^2 e^{x^1}\},$

$\qquad \mathcal{Q}(\mathcal{M}_2^6) = \mathrm{span}\{1, e^{x^2}, e^{-x^1}\}, \qquad \mathcal{Q}(\mathcal{M}_3^6) = \mathrm{span}\{1, x^1, e^{x^2}\},$

$\qquad \mathcal{Q}(\mathcal{M}_4^6) = \mathrm{span}\{1, x^2, (x^2)^2 + 2x^1\}, \quad \mathcal{Q}(\mathcal{M}_5^6) = \mathrm{span}\{1, e^{x^1}\cos(x^2), e^{x^1}\sin(x^2)\}.$

$\mathcal{Q}(\mathcal{M}_i^4)$: $\mathcal{Q}(\mathcal{M}_1^4) = \text{span}\{e^{x^2}, x^2 e^{x^2}, e^{-x^1+x^2}\}$, $\mathcal{Q}(\mathcal{M}_2^4(c_1)) = e^{c_1 x^2} \text{span}\{\mathbb{1}, e^{x^2}, e^{-x^1}\}$,

$\quad \mathcal{Q}(\mathcal{M}_3^4(c_1)) = e^{c_1 x^2} \text{span}\{\mathbb{1}, e^{x^2}, x^1\}$, $\quad \mathcal{Q}(\mathcal{M}_4^4(c)) = e^{x^2} \text{span}\{\mathbb{1}, x^2, c(x^2)^2 + 2x^1\}$,

$\quad \mathcal{Q}(\mathcal{M}_5^4(c)) = \text{span}\{e^{cx^2}\cos(x^2), e^{cx^2}\sin(x^2), e^{x^1}\}$.

$\mathcal{Q}(\mathcal{M}_i^2)$: $\mathcal{Q}(\mathcal{M}_1^2(a_1, a_2)) = \text{span}\{e^{x^1}, e^{x^2}, e^{a_1 x^1 + a_2 x^2}\}$,

$\quad \mathcal{Q}(\mathcal{M}_2^2(b_1, b_2)) = \text{span}\{e^{x^1}\cos(x^2), e^{x^1}\sin(x^2), e^{b_1 x^1 + b_2 x^2}\}$,

$\quad \mathcal{Q}(\mathcal{M}_3^2(c_2)) = e^{x^1} \text{span}\{\mathbb{1}, x^1 - c_2 x^2, e^{x^2}\}$,

$\quad \mathcal{Q}(\mathcal{M}_4^2(\pm 1)) = \text{span}\{e^{x^1}, x^2 e^{x^1}, (2x^1 \pm (x^2)^2)e^{x^1}\}$.

Proof. Our initial investigation was informed by Lemma 12.53, which shows that $\mathcal{Q}(\cdot)$ was generated by exponentials and polynomials; it was not simply routine exercise using mathematica. But once we had determined the solution space, one could then perform a direct computation to see that the functions given belong to $\mathcal{Q}(\cdot)$. As $\mathcal{M}$ is strongly projectively flat, $\dim(\mathcal{Q}(\cdot)) = 3$. Consequently, for dimensional reasons, these functions span. $\qquad \square$

Corollary 13.10 *If $\mathcal{M}_i^j(\cdot)$ is linearly equivalent to $\mathcal{M}_k^\ell(\cdot)$, then $i = k$ and $j = \ell$.*

Proof. It is immediate by inspection that $\mathcal{Q}(\mathcal{M}_i^j(\cdot))$ is not linearly isomorphic to $\mathcal{Q}(\mathcal{M}_k^\ell(\cdot))$ for $(i, j) \neq (k, \ell)$. $\qquad \square$

We note that there can be linear equivalences within these classes. For example, interchanging the roles of x^1 and x^2 interchanges $\mathcal{Q}(\mathcal{M}_1^2(a_1, a_2))$ and $\mathcal{Q}(\mathcal{M}_1^2(a_2, a_1))$, and thus $\mathcal{M}_1^2(a_1, a_2)$ and $\mathcal{M}_1^2(a_2, a_1)$ are linearly equivalent. We will study this point in more detail in Section 13.4.

13.2.4 OTHER EIGENSPACES OF THE AFFINE QUASI-EINSTEIN EQUATION.

We complexify to define $E_{\mathbb{C}}(\mu, \mathcal{M}_i^j(\cdot))$; real solutions can then be obtained by taking the real and imaginary parts. Our computations were informed by Lemma 12.53; the eigenspaces $E(\mu, \mathcal{M})$ are spanned by the product of linear exponentials and polynomials.

Theorem 13.11 *Let $\mu \neq -1$.*

1. $E(\mu, \mathcal{M}_i^6) = \mathcal{Q}(\mathcal{M}_i^6)$ *for* $0 \leq i \leq 5$.

2. $E_{\mathbb{C}}(\mu, \mathcal{M}_1^4) = \text{span}\{e^{(1+\sqrt{1+\mu})x^2}, e^{(1-\sqrt{1+\mu})x^2}\}$.

3. *Let* $D := 1 + 4c_1 + 4c_1^2 + 4c_1\mu + 4c_1^2\mu$. *Let* $\lambda_\pm := \frac{1}{2}(1 + 2c_1 \pm \sqrt{D})$.

    (a) *If* $D \neq 0$, $E_{\mathbb{C}}(\mu, \mathcal{M}_2^4(c_1)) = E_{\mathbb{C}}(\mu, \mathcal{M}_3^4(c_1)) = \text{span}\{e^{\lambda_+ x^2}, e^{\lambda_- x^2}\}$.

    (b) *If* $D = 0$, $E_{\mathbb{C}}(\mu, \mathcal{M}_2^4(c_1)) = E_{\mathbb{C}}(\mu, \mathcal{M}_3^4(c_1)) = \text{span}\{e^{\lambda_+ x^2}, x^2 e^{\lambda_+ x^2}\}$.

4. $E_{\mathbb{C}}(\mu, \mathcal{M}_4^4(c)) = \text{span}\{e^{(1+\sqrt{1+\mu})x^2}, e^{(1-\sqrt{1+\mu})x^2}\}$.

5. *Let* $D = c^2 + \mu + c^2\mu$. *Let* $\lambda_\pm := c \pm \sqrt{D}$.

(a) *If $D \neq 0$, $E_{\mathbb{C}}(\mu, \mathcal{M}_5^4(c)) = \operatorname{span}\{e^{\lambda + x^2}, e^{\lambda - x^2}\}$.*

(b) *If $D = 0$, $E_{\mathbb{C}}(\mu, \mathcal{M}_5^4(c)) = \operatorname{span}\{e^{cx^2}, x^2 e^{cx^2}\}$.*

6. $E(0, \mathcal{M}_i^2) = \operatorname{span}\{\mathbb{1}\}$. *If $\mu \neq 0$, then $E(\mu, \mathcal{M}_i^2) = \{0\}$.*

Proof. Assertion 1 is a direct consequence of Definition 13.7 since $\rho = 0$ for $\mathcal{M}_i^6$. Let $\mathcal{M} = \mathcal{M}_i^j(\cdot)$ for $j = 2, 4$. Since $\mathcal{M}$ is not flat, $\dim(E(\mu, \mathcal{M})) \neq 3$ for $\mu \neq -1$. Thus $\dim(E(\mu, \mathcal{M})) \leq 2$. We may then verify Assertions 2–5 by a direct computation. Suppose the Ricci tensor has rank 2. Let $f \in E(\mu, \mathcal{M})$ for $\mu \neq -1$. We apply Lemma 13.2 to see $R_{12}(df) = 0$. Since R_{12} is injective, this implies $df = 0$ and f is constant. Assertion 6 follows. $\qquad\square$

13.2.5 LOCAL AFFINE EQUIVALENCES.

Lemma 13.12

1. *$\Phi_1^6(x^1, x^2) := (e^{x^1}, x^2 e^{x^1})$ is an affine embedding of $\mathcal{M}_1^6$ in $\mathcal{M}_0^6$.*
2. *$\Phi_2^6(x^1, x^2) := (e^{x^2}, e^{-x^1})$ is an affine embedding of $\mathcal{M}_2^6$ in $\mathcal{M}_0^6$.*
3. *$\Phi_3^6(x^1, x^2) := (x^1, e^{x^2})$ is an affine embedding of $\mathcal{M}_3^6$ in $\mathcal{M}_0^6$.*
4. *$\Phi_4^6(x^1, x^2) := (x^2, (x^2)^2 + 2x^1)$ is an affine isomorphism from $\mathcal{M}_4^6$ to $\mathcal{M}_0^6$.*
5. *$\Phi_5^6(x^1, x^2) := (e^{x^1} \cos(x^2), e^{x^1} \sin(x^2))$ is an affine immersion of $\mathcal{M}_5^6$ in $\mathcal{M}_0^6$.*
6. *$\Phi_1^4(x^1, x^2) := (e^{-x^1}, x^2)$ is an affine embedding of $\mathcal{M}_1^4$ in $\mathcal{M}_4^4(0)$.*
7. *$\Phi_2^4(x^1, x^2) := (e^{-x^1}, x^2)$ is an affine embedding of $\mathcal{M}_2^4(c_1)$ in $\mathcal{M}_3^4(c_1)$.*
8. *$\Phi_3^4(x^1, x^2) := (x^1 e^{-x^2}, -x^2)$ is an affine isomorphism from $\mathcal{M}_3^4(c_1)$ to $\mathcal{M}_3^4(-c_1 - 1)$.*
9. *$\Phi_4^4(c)(x^1, x^2) := (x^1 + \frac{1}{2}c(x^2)^2, x^2)$ is an affine isomorphism from $\mathcal{M}_4^4(c)$ to $\mathcal{M}_4^4(0)$.*
10. *$\Phi_5^4(x^1, x^2) := (x^1, -x^2)$ is an affine isomorphism from $\mathcal{M}_5^4(c)$ to $\mathcal{M}_5^4(-c)$.*

Proof. We use Lemma 13.9 to show that the Φ_i^j intertwine the solution spaces $\mathcal{Q}$; these maps are therefore affine maps by Theorem 12.34. $\qquad\square$

Remark 13.13   We use Lemma 13.9 to obtain that

$$\mathcal{Q}(\mathcal{M}_4^6) = \operatorname{span}\{\mathbb{1}, x^2, (x^2)^2 + 2x^1\},$$
$$\mathcal{Q}(\mathcal{M}_0^6) = \operatorname{span}\{\mathbb{1}, x^1, x^2\},$$
$$\mathcal{Q}(\mathcal{M}_4^4(c)) = e^{x^2} \operatorname{span}\{\mathbb{1}, x^2, c(x^2)^2 + 2x^1\}.$$

The solution spaces $\mathcal{Q}(\mathcal{M}_0^6)$ and $\mathcal{Q}(\mathcal{M}_4^6)$ are not linearly equivalent. Consequently, by Theorem 12.34, $\mathcal{M}_0^6$ is not linearly equivalent to $\mathcal{M}_4^6$. However, these two spaces are affine equivalent by Lemma 13.12. Similarly, $\mathcal{Q}(\mathcal{M}_4^4(1))$ is not linearly equivalent to $\mathcal{Q}(\mathcal{M}_4^4(0))$, and thus

$\mathcal{M}_4^4(1)$ is not linearly equivalent to $\mathcal{M}_4^4(0)$. However, again by Lemma 13.12, there is an affine isomorphism between $\mathcal{M}_4^4(c)$ and $\mathcal{M}_4^4(0)$ for any c. Thus Theorem 13.5 fails if the Ricci tensor is not assumed to be non-degenerate.

13.2.6 AFFINE GRADIENT RICCI SOLITONS.

We say that a smooth 1-form ω is an *affine Ricci soliton* if $\nabla\omega + \rho = 0$ and that a smooth function f is an *affine gradient Ricci soliton* if df is an affine Ricci soliton or, equivalently, if $\mathcal{H}f + \rho_s = 0$. In the Type $\mathcal{A}$ setting, the Ricci tensor is symmetric. Thus $\nabla\omega = -\rho$ implies $\omega_{i;j} = \omega_{j;i} = -\rho_{ij}$. This symmetry implies that ω is closed. Consequently, since $\mathbb{R}^2$ is simply connected, ω is exact, and thus these two notions coincide in the setting at hand.

Let $\mathfrak{A}(\mathcal{M})$ be the set of all affine gradient Ricci solitons. If $f_0 \in \mathfrak{A}(\mathcal{M})$ is an affine gradient Ricci soliton and if $f_1 \in E(0,\mathcal{M})$ is a Yamabe soliton, then $f_0 + f_1 \in \mathfrak{A}(\mathcal{M})$ is again an affine gradient Ricci soliton. Thus $\mathfrak{A}(\mathcal{M})$ is an affine space. The spaces $E(0,\mathcal{M})$ of Yamabe solitons are given in Theorem 13.11. Consequently, to describe $\mathfrak{A}(\mathcal{M})$, we must either show $\mathfrak{A}(\mathcal{M})$ is empty or exhibit a single element.

Theorem 13.14   *Adopt the notation of Definition 13.7.*

1. *Let $\mathcal{M} = \mathcal{M}_i^6$ for $0 \le i \le 5$. Then $\mathbb{1} \in \mathfrak{A}(\mathcal{M})$.*
2. *Let $\mathcal{M} = \mathcal{M}_1^4$ or $\mathcal{M} = \mathcal{M}_4^4(c)$. Then $\frac{x^2}{2} \in \mathfrak{A}(\mathcal{M})$.*
3. *Let $\mathcal{M} = \mathcal{M}_2^4(c_1)$ or $\mathcal{M} = \mathcal{M}_3^4(c_1)$. If $c_1 \ne -\frac{1}{2}$, then $\frac{c_1(1+c_1)}{1+2c_1}x^2 \in \mathfrak{A}(\mathcal{M})$. If $c_1 = -\frac{1}{2}$, then $\frac{1}{8}(x^2)^2 \in \mathfrak{A}(\mathcal{M})$.*
4. *Let $\mathcal{M} = \mathcal{M}_5^4(c)$. If $c \ne 0$, then $\frac{1+c^2}{2c}x^2 \in \mathfrak{A}(\mathcal{M})$. If $c = 0$, then $-\frac{1}{2}(x^2)^2 \in \mathfrak{A}(\mathcal{M})$.*
5. *If $\mathcal{M} = \mathcal{M}_i^2(\cdot)$, $1 \le i \le 4$, then $\mathfrak{A}(\mathcal{M})$ is empty.*

Proof. A direct calculation establishes the first four assertions. Suppose that $\mathcal{M}$ is a Type $\mathcal{A}$ model where the Ricci tensor is non-degenerate. By Lemma 13.2, $R_{12}(df) = 0$. Since ρ is non-degenerate, this implies $df = 0$. This shows that $\mathcal{H}f = 0$. Thus it is not possible that one has $\mathcal{H}f = -\rho_s$. This proves Assertion 5. $\qquad\square$

13.2.7 AFFINE KILLING VECTOR FIELDS.

Recall that $\mathcal{M}$ is said to be *affine Killing complete* if every affine Killing vector field of $\mathcal{M}$ is a complete vector field, i.e., the integral curves exist for all time or, equivalently by Lemma 12.3, that $\mathfrak{K}(\mathcal{M})$ is the Lie algebra of the Lie group of affine diffeomorphisms $\mathrm{Aff}(\mathcal{M})$. We refer to Nomizu [48]. We divide our analysis into three cases depending upon the rank of the Ricci tensor. Our initial investigation was informed by Lemma 12.53, which shows that $\mathfrak{K}(\cdot)$ is generated by exponentials and polynomials; the results of this section are not simply routine exercise using mathematica. But once we had determined a sufficient number of affine Killing vector fields, we then proved they spanned for dimensional reasons by Lemma 13.4. We first study the flat geometries; the affine plane is affine Killing complete and is modeled on any of these geometries.

Lemma 13.15   *The geometries $\mathcal{M}_i^6$ are all flat. Consequently, $\dim(\mathfrak{K}(\mathcal{M}_i^6)) = 6$.*

1. $\mathfrak{K}(\mathcal{M}_0^6) = \mathrm{span}\{\partial_{x^1}, x^2\partial_{x^1}, x^1\partial_{x^1}, \partial_{x^2}, x^1\partial_{x^2}, x^2\partial_{x^2}\}$.

2. $\mathfrak{K}(\mathcal{M}_1^6) = \mathrm{span}\{\partial_{x^1}, \partial_{x^2}, e^{-x^1}(\partial_{x^1} - x^2\partial_{x^2}), x^2\partial_{x^1} - (x^2)^2\partial_{x^2}, x^2\partial_{x^2}, e^{-x^1}\partial_{x^2}\}$.

3. $\mathfrak{K}(\mathcal{M}_2^6) = \mathrm{span}\{\partial_{x^1}, \partial_{x^2}, e^{x^1}\partial_{x^1}, e^{x^1+x^2}\partial_{x^1}, e^{-x^2}\partial_{x^2}, e^{-x^1-x^2}\partial_{x^2}\}$.

4. $\mathfrak{K}(\mathcal{M}_3^6) = \mathrm{span}\{\partial_{x^1}, \partial_{x^2}, x^1\partial_{x^1}, e^{x^2}\partial_{x^1}, e^{-x^2}\partial_{x^2}, x^1e^{-x^2}\partial_{x^2}\}$.

5. $\mathfrak{K}(\mathcal{M}_4^6) = \mathrm{span}\{\partial_{x^1}, \partial_{x^2}, x^2\partial_{x^1}, (-x^1x^2 - \frac{1}{2}(x^2)^3)\partial_{x^1} + (x^1 + \frac{1}{2}(x^2)^2)\partial_{x^2},$
$(x^1 + \frac{1}{2}(x^2)^2)\partial_{x^1}, -(x^2)^2\partial_{x^1} + x^2\partial_{x^2}\}$.

6. $\mathfrak{K}(\mathcal{M}_5^6) = \mathrm{span}\{\partial_{x^1}, \partial_{x^2}, \cos(2x^2)\partial_{x^1} - \sin(2x^2)\partial_{x^2}, \sin(2x^2)\partial_{x^1} + \cos(2x^2)\partial_{x^2},$
$e^{-x^1}(\cos(x^2)\partial_{x^1} - \sin(x^2)\partial_{x^2}), e^{-x^1}(\sin(x^2)\partial_{x^1} + \cos(x^2)\partial_{x^2})\}$.

7. $\mathcal{M}_0^6$ *and* $\mathcal{M}_4^6$ *are affine Killing complete;* $\mathcal{M}_1^6$, $\mathcal{M}_2^6$, $\mathcal{M}_3^6$, *and* $\mathcal{M}_5^6$ *are affine Killing incomplete. All these Lie algebras are isomorphic to the full Lie algebra of the 6-dimensional affine group.*

Proof. Assertions 1–6 follow by a direct computation. We have $\mathrm{Aff}(\mathcal{M}_0^6)$ is the 6-dimensional affine group. Thus, for dimensional reasons, the Lie algebra of $\mathrm{Aff}(\mathcal{M}_0^6)$ is the full Lie algebra of affine Killing vector fields and $\mathcal{M}_0^6$ is affine Killing complete. By Lemma 13.12, $\mathcal{M}_4^6$ is affine diffeomorphic to $\mathcal{M}_0^6$. This shows that $\mathcal{M}_4^6$ is affine Killing complete. The remaining geometries $\mathcal{M}_i^6$ all have proper affine embeddings or immersions into $\mathcal{M}_0^6$. It then follows that these must be affine Killing incomplete. It is instructive, however, to give a direct argument.

Case 1. $\mathcal{M}_1^6$. Let $X = x^2\partial_{x^1} - (x^2)^2\partial_{x^2}$. The flow is determined by the ODE $\dot{x}^1 = x^2$ and $\dot{x}^2 = -(x^2)^2$. We solve this ODE to see $x^2(t) = \frac{1}{t+a}$ so the integral curve is not defined for all t and $\mathcal{M}_1^6$ is affine Killing incomplete.

Case 2. $\mathcal{M}_2^6$. Let $X = e^{x^1}\partial_{x^1}$. The flow is determined by the ODE $\dot{x}^1 = e^{x^1}$ and $\dot{x}^2 = 0$. The first ODE can be solved by setting $x^1(t) = -\log(-t + a)$ so the integral curve is not defined for all t and $\mathcal{M}_2^6$ is affine Killing incomplete.

Case 3. $\mathcal{M}_3^6$. Let $X = e^{-x^2}\partial_{x^2}$. The flow is determined by the ODE $\dot{x}^1 = 0$ and $\dot{x}^2 = e^{-x^2}$. We obtain $x^2(t) = \log(t + a)$ so the integral curve is not defined for all t and $\mathcal{M}_3^6$ is affine Killing incomplete.

Case 4. $\mathcal{M}_5^6$. Let $X = e^{-x^1}\cos(x^2)\partial_{x^1} - e^{-x^1}\sin(x^2)\partial_{x^2}$. The flow is determined by the ODE $\dot{x}^1 = e^{-x^1}\cos(x^2)$ and $\dot{x}^2 = -e^{-x^1}\sin(x^2)$. We set $x^2(t) = 0$ to solve the second ODE. The first ODE then becomes $\dot{x}^1 = e^{-x^1}$, which yields $x^1(t) = \log(t + a)$ so the integral curve is not defined for all t and $\mathcal{M}_5^6$ is affine Killing incomplete. $\qquad\square$

Lemma 13.16   *The geometries* $\mathcal{M}_i^4(\cdot)$ *satisfy* $\dim(\mathfrak{K}(\mathcal{M}_i^4(\cdot))) = 4$.

1. $\mathfrak{K}(\mathcal{M}_1^4) = \mathrm{span}\{\partial_{x^1}, \partial_{x^2}, e^{x^1}\partial_{x^1}, x^2 e^{x^1}\partial_{x^1}\}$.

2. $\mathfrak{K}(\mathcal{M}_2^4(c_1)) = \mathrm{span}\{\partial_{x^1}, \partial_{x^2}, e^{x^1}\partial_{x^1}, e^{x^1+x^2}\partial_{x^1}\}$.

3. $\mathfrak{K}(\mathcal{M}_3^4(c_1)) = \mathrm{span}\{\partial_{x^1}, \partial_{x^2}, e^{x^2}\partial_{x^1}, x^1\partial_{x^1}\}$.

4. $\mathfrak{K}(\mathcal{M}_4^4(c)) = \mathrm{span}\{\partial_{x^1}, \partial_{x^2}, (x^1 + \frac{1}{2}c(x^2)^2)\partial_{x^1}, x^2\partial_{x^1}\}$.

5. $\mathfrak{K}(\mathcal{M}_5^4(c)) = \mathrm{span}\{\partial_{x^1}, \partial_{x^2}, e^{-x^1+cx^2}\cos(x^2)\partial_{x^1}, e^{-x^1+cx^2}\sin(x^2)\partial_{x^1}\}$.

6. *The geometries* $\mathcal{M}_3^4(c_1)$ *and* $\mathcal{M}_4^4(c)$ *are affine Killing complete.*

7. *The geometries* $\mathcal{M}_1^4$, $\mathcal{M}_2^4(c_1)$, *and* $\mathcal{M}_5^4(c)$ *are affine Killing incomplete.*

Proof. By Lemma 13.8, the Ricci tensor has rank 1. Lemma 13.4 yields $\dim(\mathfrak{K}(\cdot)) \leq 4$. We perform a direct computation to verify that the vector fields given are all affine Killing vector fields and thus for dimensional reasons span $\mathfrak{K}(\cdot)$. Assertions 1–5 now follow. We break the proof of Assertion 6 into two cases.

Case 1. $\mathcal{M} = \mathcal{M}_3^4(c_1)$. We have $\mathcal{Q}(\mathcal{M}) = \mathrm{span}\{e^{c_1 x^2}, e^{(1+c_1)x^2}, x^1 e^{c_1 x^2}\}$. This is not a particularly convenient form of this surface to work with. We set $u^1 := x^1 e^{c_1 x^2}$ and $u^2 := x^2$ to express $\mathcal{Q}(\mathcal{M}_3^4(c_1)) = \mathrm{span}\{e^{c_1 u^2}, e^{(1+c_1)u^2}, u^1\}$. We define

$$T(\alpha, \beta, \gamma, \delta)(u^1, u^2) = (e^\alpha u^1 + \beta e^{c_1 u^2} + \gamma e^{(1+c_1)u^2}, u^2 + \delta).$$

Because $T(\alpha, \beta, \gamma, \delta)$ preserves $\mathcal{Q}(\mathcal{M})$, $T(\alpha, \beta, \gamma, \delta) \in \mathrm{Aff}(\mathcal{M})$. The set of these elements is closed under composition and inverse:

$$T(\alpha, \beta, \gamma, \delta) \circ T(\tilde{\alpha}, \tilde{\beta}, \tilde{\gamma}, \tilde{\delta}) = T(\alpha + \tilde{\alpha}, \tilde{\beta}e^\alpha + \beta e^{c\tilde{\delta}}, \tilde{\gamma}e^\alpha + \gamma e^{(1+c)\tilde{\delta}}, \delta + \tilde{\delta}),$$
$$T(\alpha, \beta, \gamma, \delta)^{-1} = T(-\alpha, -\beta e^{-\alpha-c\delta}, -\gamma e^{-\alpha+(-1-c)\delta}, -\delta).$$

This gives a group structure to $\mathbb{R}^4$ and identifies $\mathbb{R}^4$ with the connected component of the identity of the group $\mathrm{Aff}(\mathcal{M})$. The Lie algebra of this group is 4-dimensional and consists of the full Lie algebra of Killing vector fields. This shows that $\mathcal{M}$ is affine Killing complete. This action is transitive on $\mathbb{R}^2$ so this is a homogeneous geometry.

Case 2. By Lemma 13.12, $\mathcal{M}_4^4(c)$ is affine equivalent to $\mathcal{M}_4^4(0)$ for any c so it is only necessary to consider $\mathcal{M} = \mathcal{M}_4^4(0)$. We have $\mathcal{Q}(\mathcal{M}) = \mathrm{span}\{e^{x^2}, x^2 e^{x^2}, x^1 e^{x^2}\}$. We clear the previous notation and set $T(\alpha, \beta, \gamma, \delta)(x^1, x^2) := (e^\alpha x^1 + \beta x^2 + \gamma, x^2 + \delta)$. As in Case 1, $T(\alpha, \beta, \gamma, \delta)$ belongs to $\mathrm{Aff}(\mathcal{M})$. We complete the analysis by checking

$$T(\alpha, \beta, \gamma, \delta) \circ T(\tilde{\alpha}, \tilde{\beta}, \tilde{\gamma}, \tilde{\delta}) = T(\alpha + \tilde{\alpha}, \beta + \tilde{\beta}e^\alpha, \gamma + \beta\tilde{\delta} + \tilde{\gamma}e^\alpha, \delta + \tilde{\delta}),$$
$$T(\alpha, \beta, \gamma, \delta)^{-1} = T(-\alpha, -\beta e^{-\alpha}, e^{-\alpha}(\beta\delta - \gamma), -\delta).$$

We perform a direct computation to show $\mathcal{M}_1^4$, $\mathcal{M}_2^4(c_1)$, and $\mathcal{M}_5^4(c)$ are affine Killing incomplete, although this also follows from Lemma 13.12. If $\mathcal{M} = \mathcal{M}_1^4$ or $\mathcal{M} = \mathcal{M}_2^4(c_1)$, then

$e^{x^1}\partial_{x^1}$ is an affine Killing vector field. We noted in the proof of Lemma 13.15 (see Case 2) that this vector field was incomplete. If $\mathcal{M} = \mathcal{M}_5^4(c)$, we take $X = e^{-x^1 + cx^2}\cos(x^2)\partial_{x^1}$. We take $x^2 = 0$ to obtain $X = e^{-x^1}\partial_{x^1}$. We showed in the proof of Lemma 13.15 (see Case 3 and interchange the roles of x^1 and x^2) that this vector field was incomplete.   □

Patera et al. [55] have classified the low-dimensional Lie algebras. Let $\{e_1, e_2, e_3, e_4\}$ be a basis of $\mathbb{R}^4$. We define the following solvable Lie algebras by specifying their bracket relations.

1. $A_2 \oplus A_2$: the relations of the bracket are given by $[e_1, e_2] = e_2$ and $[e_3, e_4] = e_4$.

2. $A_{4,9}^b$: the relations of the bracket for $-1 \leq b \leq 1$ are given by $[e_2, e_3] = e_1$, $[e_1, e_4] = (1+b)e_1$, $[e_2, e_4] = e_2$, and $[e_3, e_4] = be_3$.

3. $A_{4,12}$: the relations of the bracket are given by $[e_1, e_3] = e_1$, $[e_2, e_3] = e_2$, $[e_1, e_4] = -e_2$, $[e_2, e_4] = e_1$.

Lemma 13.17

1. $\mathfrak{K}(\mathcal{M}_1^4) \approx \mathfrak{K}(\mathcal{M}_4^4(c)) \approx A_{4,9}^0$.

2. $\mathfrak{K}(\mathcal{M}_2^4(c_1)) \approx \mathfrak{K}(\mathcal{M}_3^4(c_1)) \approx A_2 \oplus A_2$.

3. $\mathfrak{K}(\mathcal{M}_5^4(c)) \approx A_{4,12}$.

Proof. In view of the embeddings and isomorphisms of Lemma 13.12, we need only consider three cases. We follow the discussion of Brozos-Vázquez, García-Río, and Gilkey [7].

Case 1. $\mathcal{M} = \mathcal{M}_4^4(0)$. Set $e_1 := \partial_{x^1}$, $e_2 := x^2\partial_{x^1}$, $e_3 := -\partial_{x^2}$, $e_4 := x^1\partial_{x^1}$ to obtain the bracket relations of the Lie algebra $A_{4,9}^0$: $[e_2, e_3] = e_1$, $[e_1, e_4] = e_1$, and $[e_2, e_4] = e_2$.

Case 2. $\mathcal{M} = \mathcal{M}_3^4(c_1)$. Set $e_1 := -x^1\partial_{x^1} - \partial_{x^2}$, $e_2 := -\partial_{x^1}$, $e_3 := \partial_{x^2}$, and $e_4 := e^{x^2}\partial_{x^1}$ to obtain the bracket relations of the Lie algebra $A_2 \oplus A_2$: $[e_1, e_2] = e_2$ and $[e_3, e_4] = e_4$.

Case 3. $\mathcal{M} = \mathcal{M}_5^4(c)$. Set $e_1 := e^{x^1}\cos(x^2)\partial_{x^1}$, $e_2 := e^{x^1}\sin(x^2)\partial_{x^1}$, $e_3 := -\partial_{x^1}$, and $e_4 := -\partial_{x^2}$, to obtain the bracket relations of $A_{4,12}$: $[e_1, e_3] = e_1$, $[e_2, e_3] = e_2$, $[e_1, e_4] = -e_2$, and $[e_2, e_4] = e_1$.   □

By Lemma 13.12, the map $(x^1, x^2) \rightarrow (e^{-x^1}, x^2)$ is a proper affine embedding of $\mathcal{M}_1^4$ in $\mathcal{M}_4^4(0)$ and of $\mathcal{M}_2^4(c_1)$ in $\mathcal{M}_3^4(c_1)$. Thus these geometries can be affine Killing completed. We now introduce geometries $\tilde{\mathcal{M}}_5^4(c)$ which contain $\mathcal{M}_5^4(c)$ as a proper submanifold and which can be regarded as a completion of the geometries $\mathcal{M}_5^4(c)$.

Lemma 13.18   Let $\tilde{\mathcal{M}}_5^4(c) := (\mathbb{R}^2, \tilde{\nabla})$ where the only (possibly) non-zero Christoffel symbols of $\tilde{\nabla}$ are $\tilde{\Gamma}_{22}^1 = (1 + c^2)x^1$ and $\tilde{\Gamma}_{22}^2 = 2c$.

1. $\mathcal{Q}(\tilde{\mathcal{M}}_5^4(c)) = \text{span}\{e^{cx^2}\cos(x^2), e^{cx^2}\sin(x^2), x^1\}$.

2. $\tilde{\mathcal{M}}_5^4(c)$ is a homogeneous geometry which is affine Killing complete.

3. The map $\Theta_5^4(x^1, x^2) := (e^{x^1}, x^2)$ is an affine embedding of $\mathcal{M}_5^4(c)$ in $\tilde{\mathcal{M}}_5^4(c)$.

4. $\tilde{\mathcal{M}}_5^4(c)$ *is an affine symmetric space if and only if $c = 0$.*

Proof. We make a direct computation to see

$$\{e^{cx^2} \cos(x^2), e^{cx^2} \sin(x^2), x^1\} \subset \mathcal{Q}(\tilde{\mathcal{M}}_5^4(c)).$$

Since $\dim(\mathcal{Q}(\tilde{\mathcal{M}}_5^4(c)) \le 3$, Assertion 1 holds for dimensional reasons. Let

$$T(\alpha, \beta, \gamma, \delta)(x^1, x^2) := (e^\alpha x^1 + \beta e^{cx^2} \cos(x^2) + \gamma e^{cx^2} \sin(x^2), x^2 + \delta).$$

Since $T(\alpha, \beta, \gamma, \delta)$ preserves $\mathcal{Q}(\tilde{\mathcal{M}}_5^4(c))$, $T(\alpha, \beta, \gamma, \delta) \in \mathrm{Aff}(\tilde{\mathcal{M}}_5^4(c))$. The set of these elements is closed under composition and inverse:

$$T(\alpha, \beta, \gamma, \delta) \circ T(\tilde{\alpha}, \tilde{\beta}, \tilde{\gamma}, \tilde{\delta})$$
$$= T(\alpha + \tilde{\alpha}, e^\alpha \tilde{\beta} + \beta e^{c\tilde{\delta}} \cos(\tilde{\delta}) + \gamma e^{c\tilde{\delta}} \sin(\tilde{\delta}), e^\alpha \tilde{\gamma} - \beta e^{c\tilde{\delta}} \sin(\tilde{\delta}) + \gamma e^{c\tilde{\delta}} \cos(\tilde{\delta}), \delta + \tilde{\delta}),$$
$$T(\alpha, \beta, \gamma, \delta)^{-1}$$
$$= T(-\alpha, -e^{-\alpha - c\delta}(\beta \cos(\delta) - \gamma \sin(\delta)), -e^{-\alpha - c\delta}(\beta \sin(\delta) + \gamma \cos(\delta)), -\delta).$$

We argue as in the proof of Lemma 13.16 to see that $\tilde{\mathcal{M}}_5^4(c))$ is affine Killing complete. This action is transitive on $\mathbb{R}^2$ so this is a homogeneous geometry. This establishes Assertion 2. Assertion 3 follows since Θ_5^4 intertwines $\mathcal{Q}(\mathcal{M}_5^4(c))$ and $\mathcal{Q}(\tilde{\mathcal{M}}_5^4(c))$. Assertion 4 follows from the corresponding assertion for $\mathcal{M}_5^4(c)$ given in Lemma 13.8. □

Lemma 13.19 *Let $\mathcal{M} = \mathcal{M}_i^2(\cdot)$ for $1 \le i \le 4$. Then $\mathfrak{K}(\mathcal{M}) = \mathrm{span}\{\partial_{x^1}, \partial_{x^2}\}$ and $\mathcal{M}$ is affine Killing complete.*

Proof. By Lemma 13.8, $\mathrm{Rank}(\rho_{\mathcal{M}_i^2(\cdot)}) = 2$ and thus by Lemma 13.4, $\dim(\mathfrak{K}(\mathcal{M}_i^2(\cdot))) = 2$. Thus $\mathfrak{K}(\mathcal{M}_i^2(\cdot)) = \mathrm{span}\{\partial_{x^1}, \partial_{x^2}\}$. Since the coordinate vector fields are complete, $\mathcal{M}_i^2(\cdot)$ is affine Killing complete. □

13.2.8 TYPE (1,1) INVARIANT AND PARALLEL TENSORS ON TYPE $\mathcal{A}$ SURFACES.

Let $\mathcal{M} = (\mathbb{R}^2, \nabla)$ be a Type $\mathcal{A}$ structure which is not flat. Suppose given a tensor of Type (1,1) $T = T^i{}_j \partial_{x^i} \otimes dx^j$, i.e., an endomorphism of the tangent bundle. We assume T is an *invariant* tensor of Type (1,1), i.e., $\mathcal{L}_X T = 0$ for all $X \in \mathfrak{K}(\mathcal{M})$.

Theorem 13.20 *Let $T = T^i{}_j(x^1, x^2)$ be a tensor of Type (1,1).*

1. *If $\mathcal{M} = \mathcal{M}_i^2$, then T is invariant if and only if the component functions $T^i{}_j$ are constant.*

2. *If $\mathcal{M} = \mathcal{M}_i^6$ or if $\mathcal{M} = \mathcal{M}_i^4$, then T is invariant if and only if T is a constant multiple of the identity.*

Proof. We have ∂_{x^1} and ∂_{x^2} are affine Killing vector fields. Since $\mathcal{L}_{\partial_{x^i}}\partial_{x^j} = 0$, we have dually that $\mathcal{L}_{\partial_{x^i}}dx^j = 0$. Consequently, $\mathcal{L}_{\partial_{x^i}}(T^a{}_b\partial_{x^a} \otimes dx^b) = (\partial_{x^i}T^a{}_b)\partial_{x^a} \otimes dx^b$. Thus T is invariant implies $T \in M_2(\mathbb{R})$ is constant. If $\dim(\mathfrak{K}(\mathcal{M})) = 2$, this implies T is invariant; Assertion 1 now follows. More generally, let $X = a^1\partial_{x^1} + a^2\partial_{x^2}$. By Equation (12.2.a), $\mathcal{L}_X T = 0$ is equivalent to the conditions

$$0 = -T^k{}_j(\partial_{x^k}a^i) + T^i{}_k(\partial_{x^j}a^k) \quad \text{for all} \quad i, j .$$

Clearly if T is a constant multiple of the identity, then $\mathcal{L}_X T = 0$. Thus we may assume T is tracefree henceforth, i.e., $T^1{}_1 + T^2{}_2 = 0$. We impose this condition. Then $\mathcal{L}_X T = 0$ is equivalent to the conditions

$$\begin{aligned}
0 &= \{\mathcal{L}_X T\}^1{}_1 = T^1{}_2(a^2)^{(1,0)} - T^2{}_1(a^1)^{(0,1)}, \\
0 &= \{\mathcal{L}_X T\}^1{}_2 = T^1{}_2\{(a^2)^{(0,1)} - (a^1)^{(1,0)}\} + 2T^1{}_1(a^1)^{(0,1)}, \\
0 &= \{\mathcal{L}_X T\}^2{}_1 = T^2{}_1(a^1)^{(1,0)} - 2T^1{}_1(a^2)^{(1,0)} - T^2{}_1(a^2)^{(0,1)}, \\
0 &= \{\mathcal{L}_X T\}^2{}_2 = T^2{}_1(a^1)^{(0,1)} - T^1{}_2(a^2)^{(1,0)} .
\end{aligned}$$

We apply Lemma 13.12 to see that we need only consider the geometries $\mathcal{M}_0^6$, $\mathcal{M}_3^4(c_1)$, $\mathcal{M}_4^4(0)$, and $\mathcal{M}_5^4(c)$. We consider cases. We apply the results of Section 13.2.7 to determine the Lie algebra of affine Killing vector fields.

Case 1. $\mathcal{M} = \mathcal{M}_0^6$. We have $X = (b_{11}x^1 + b_{12}x^2)\partial_{x^1} + (b_{21}x^1 + b_{22}x^2)\partial_{x^2}$ is an affine Killing vector field. We set $T^2{}_2 = -T^1{}_1$ and compute:

$$\mathcal{L}_X T = \begin{pmatrix} b_{21}T^1{}_2 - b_{12}T^2{}_1 & 2b_{12}T^1{}_1 + (b_{22} - b_{11})T^1{}_2 \\ (b_{11} - b_{22})T^2{}_1 - 2b_{21}T^1{}_1 & b_{12}T^2{}_1 - b_{21}T^1{}_2 \end{pmatrix} .$$

Since this must vanish identically for all values of the parameters b_{ij}, we conclude $T = 0$.

Case 2. $\mathcal{M} = \mathcal{M}_3^4(c_1)$. We have $X = (b_1 e^{x^2} + b_2 x^1)\partial_{x^1}$ is an affine Killing vector field. We compute

$$\mathcal{L}_X T = \begin{pmatrix} -b_1 e^{x^2}T^2{}_1 & 2b_1 e^{x^2}T^1{}_1 - b_2 T^1{}_2 \\ b_2 T^2{}_1 & b_1 e^{x^2}T^2{}_1 \end{pmatrix} .$$

This vanishes identically for all (b_1, b_2) if and only if $T = 0$.

Case 3. $\mathcal{M} = \mathcal{M}_4^4(c)$. We have $X = \{b_1(x^1 + \frac{1}{2}c(x^2)^2) + b_2 x^2\}\partial_{x^1}$ is an affine Killing vector field. We compute

$$\mathcal{L}_X T = \begin{pmatrix} -T^2{}_1(b_2 + b_1 c x^2) & 2T^1{}_1(b_2 + b_1 c x^2) - b_1 T^1{}_2 \\ b_1 T^2{}_1 & T^2{}_1(b_2 + b_1 c x^2) \end{pmatrix} .$$

This vanishes identically for all (b_1, b_2) if and only if $T = 0$.

Case 4. $\mathcal{M} = \mathcal{M}_5^4(c)$. We have $X = e^{-x^1+cx^2}\{b_1\cos(x^2) + b_2\sin(x^2)\}\partial_{x^1}$ is an affine Killing vector field. We compute

$$(\mathcal{L}_X T)_{x^1=0,x^2=0} = \begin{pmatrix} -(b_2 + b_1 c)T^2{}_1 & 2(b_2 + b_1 c)T^1{}_1 + b_1 T^1{}_2 \\ -b_1 T^2{}_1 & (b_2 + b_1 c)T^2{}_1 \end{pmatrix}.$$

This vanishes identically for all (b_1, b_2) if and only if $T = 0$. □

The Ricci tensor of any Type $\mathcal{A}$ homogeneous model is symmetric. Furthermore, the Ricci tensor is recurrent if and only if it is of rank 1. Therefore Theorem 12.27 3 shows that a Type $\mathcal{A}$ homogeneous surface admits a parallel tensor field if and only if the Ricci tensor is of rank 1, in which case it is a nilpotent Kähler surface.

Theorem 13.21   *Let* $\mathcal{M} = (\mathbb{R}^2, \nabla)$ *be a Type $\mathcal{A}$ model. The following assertions are equivalent.*
(1) $\text{Rank}(\rho) = 1$.   (2) $\mathcal{P}^0(\mathcal{M}) \neq \{0\}$.   (3) $\dim(\mathcal{P}^0(\mathcal{M})) = 1$.   (4) $\dim(\mathfrak{K}(\mathcal{M})) = 4$.

13.3   TYPE $\mathcal{A}$: PARAMETERIZATION

In Section 13.3.1, we parameterize the Type $\mathcal{A}$ models (see Theorem 13.22). In the remainder of this section, we exploit this classification. Section 13.3.2 relates the rank of the Ricci tensor and the dimension of the space of affine Killing vector fields. Section 13.3.3 treats strong projective equivalence and linear equivalence. The α invariant is computed for the examples where the Ricci tensor has rank 1 in Section 13.3.4. The geodesic equation is explicitly solved for the examples where the Ricci tensor has rank 0 or rank 1 and geodesic completeness is discussed in Section 13.3.5.

13.3.1 CLASSIFICATION.   Gilkey and Valle-Regueiro [35] classified the Type $\mathcal{A}$ structures up to linear equivalence using the space of solutions of the affine quasi-Einstein equation $\mathcal{Q}$; a similar classification had been given previously by Brozos-Vázquez, García-Río, and Gilkey [6] using different methods. The following is the main result of this section.

Theorem 13.22   *If* $\mathcal{M} = (\mathbb{R}^2, \nabla)$ *is a Type $\mathcal{A}$ model, then $\mathcal{M}$ is linearly equivalent to one of the structures given in Definition 13.7.*

We say that a function $L(x^1, x^2)$ is linear if $L(x^1, x^2) = a_1 x^1 + a_2 x^2$; as we permit $(a^1, a^2) = (0, 0)$, the zero function is permitted as a degenerate case. We say that a function $Q(x^1, x^2)$ is quadratic if $Q(x^1, x^2) = a_{11}(x^1)^2 + a_{12}x^1 x^2 + a_{22}(x^2)^2 + a_1 x^1 + a_2 x^2$; as we permit $(a_{11}, a_{12}, a_{22}) = (0, 0, 0)$, linear functions are permitted as a degenerate case.

Lemma 13.23   *Let $\mathcal{M} = (\mathbb{R}^2, \nabla)$ be a Type $\mathcal{A}$ model. There exists a basis $\mathcal{B}$ for $\mathcal{Q}(\mathcal{M})$ which has one of the following forms.*

1. *$\mathcal{B} = \{e^{L_1}\cos(L_2), e^{L_1}\sin(L_2), e^{L_3}\}$ where L_i are linear functions.*
2. *$\mathcal{B} = \{e^{L_1}, e^{L_2}, e^{L_3}\}$ where L_i are linear functions.*
3. *$\mathcal{B} = \{e^{L_1}, L_2 e^{L_1}, e^{L_3}\}$ where L_i are linear functions.*
4. *$\mathcal{B} = \{e^{L_1}, L_2 e^{L_1}, Q e^{L_1}\}$ where L_i are linear functions and Q is quadratic.*

Proof. We complexify and use Lemma 12.53 to see that $\mathcal{Q}_{\mathbb{C}}(\mathcal{M}) := \mathcal{Q}(\mathcal{M}) \otimes_{\mathbb{R}} \mathbb{C}$ is spanned by functions of the form $p(x^1, x^2)e^{a_1 x^1 + a_2 x^2}$ for $(a_1, a_2) \in \mathbb{C}^2$ and p polynomial. We have that $\mathcal{Q}_{\mathbb{C}}(\mathcal{M})$ is a $\mathfrak{K}(\mathcal{M})$-module. By applying $(\partial_{x^i} - a_i)$, we see that $\partial_{x^i} p(x^1, x^2)e^{a_1 x^1 + a_2 x^2}$ belongs to $\mathcal{Q}_{\mathbb{C}}(\mathcal{M})$ as well and thus inductively $e^{a_1 x^1 + a_2 x^2} \in \mathcal{Q}_{\mathbb{C}}(\mathcal{M})$. We consider cases.

Case 1. Suppose that $e^{a_1 x^1 + a_2 x^2} \in \mathcal{Q}_{\mathbb{C}}(\mathcal{M})$ where $(a_1, a_2) \notin \mathbb{R}^2$. Define real functions by setting $L_1 = \Re(a_1 x^1 + a_2 x^2)$ and $L_2 = \Im(a_1 x^1 + a_2 x^2)$. As the affine quasi-Einstein operator $\mathfrak{Q}$ is real, we may take the real and imaginary parts of $e^{a_1 x^1 + a_2 x^2}$ to see that $e^{L_1}\cos(L_2) \in \mathcal{Q}(\mathcal{M})$ and $e^{L_1}\sin(L_2) \in \mathcal{Q}(\mathcal{M})$. We now search for the remaining generator of $\mathcal{Q}(\mathcal{M})$. Suppose that there is a non-constant polynomial $p(x^1, x^2)$ so that $p(x^1, x^2)e^{a_1 x^1 + a_2 x^2} \in \mathcal{Q}_{\mathbb{C}}(\mathcal{M})$. Then

$$\{\Re(pe^{a_1 x^1 + a_2 x^2}), \Im(pe^{a_1 x^1 + a_2 x^2}), \Re(e^{a_1 x^1 + a_2 x^2}), \Im(pe^{a_1 x^1 + a_2 x^2})\} \subset \mathcal{Q}(\mathcal{M}).$$

This is impossible as $\dim(\mathcal{Q}(\mathcal{M})) = 3$. Suppose next that the missing generator takes the form $p(x^1, x^2)e^{b_1 x^1 + b_2 x^2}$ for $(b_1, b_2) \neq (a_1, a_2)$. Since $e^{b_1 x^1 + b_2 x^2} \in \mathcal{Q}_{\mathbb{C}}(\mathcal{M})$, p must be a polynomial of degree 0 and $b_1 x^1 + b_2 x^2$ is a real linear function. This is the possibility of Assertion 1.

Case 2. We suppose that all the L_i are real and that there are no polynomial factors. This is the possibility of Assertion 2.

Case 3. We suppose that all the L_i are real and there is a single polynomial factor. There must be two distinct exponentials defined by linear functions and the polynomial factor must be linear for dimensional reasons and the possibility of Assertion 3 holds.

Case 4. There is a single exponential defined by a linear function which is multiplied by polynomial factors. Two of the polynomial factors can be linear. Or one can be linear and the other quadratic; the possibility of Assertion 4 then holds.   □

We complete the proof of Theorem 13.22 by examining the possibilities given by Lemma 13.23.

Lemma 13.24   *Let $\mathcal{M}$ be a Type $\mathcal{A}$ model.*

1. *Suppose that $\mathcal{Q}(\mathcal{M}) = \mathrm{span}\{e^{L_1}\cos(L_2), e^{L_1}\sin(L_2), e^{L_3}\}$. Then $\mathcal{M}$ is linearly equivalent to $\mathcal{M}_5^6$, to $\mathcal{M}_5^4(c)$, or to $\mathcal{M}_2^2(b_1, b_2)$.*
2. *Suppose that $\mathcal{Q}(\mathcal{M}) = \mathrm{span}\{e^{L_1}, e^{L_2}, e^{L_3}\}$. Then $\mathcal{M}$ is linearly equivalent to $\mathcal{M}_2^6$, to $\mathcal{M}_2^4(c_1)$, or to $\mathcal{M}_1^2(a_1, a_2)$.*

3. *Suppose that $Q(\mathcal{M}) = \mathrm{span}\{e^{L_1}, L_2 e^{L_1}, e^{L_3}\}$. Then $\mathcal{M}$ is linearly equivalent to $\mathcal{M}_3^2(c_2)$, to $\mathcal{M}_1^4$, to $\mathcal{M}_3^4(c_1)$, to $\mathcal{M}_1^6$, or to $\mathcal{M}_3^6$.*

4. *Suppose that $Q(\mathcal{M}) = \mathrm{span}\{e^{L_1}, L_2 e^{L_1}, Q e^{L_1}\}$. Then $\mathcal{M}$ is linearly equivalent to* $\mathcal{M}_4^2(\pm 1)$, *to* $\mathcal{M}_4^4(c)$, *to* $\mathcal{M}_0^6$, *or to* $\mathcal{M}_4^6$.

Proof. We apply Theorem 12.34 to see that $Q(\mathcal{M})$ determines $\mathcal{M}$ throughout the proof.

Case 1. Suppose that $Q(\mathcal{M}) = \mathrm{span}\{e^{L_1}\cos(L_2), e^{L_1}\sin(L_2), e^{L_3}\}$. Since L_2 must be nontrivial, we can make a linear change of coordinates to assume $L_2 = x^2$.

Case 1.a. Suppose first that L_1 is not a multiple of L_2. Change coordinates to set $L_1 = x^1$ and $L_3 = b_1 x^1 + b_2 x^2$. Then $Q(\mathcal{M}) = \mathrm{span}\{e^{x^1}\cos(x^2), e^{x^1}\sin(x^2), e^{b_1 x^1 + b_2 x^2}\}$. By Lemma 12.33, $b_1 \neq 1$.

   1.1. If $(b_1, b_2) \neq (0,0)$, then we obtain $\mathcal{M}_2^2(b_1, b_2)$.

   1.2. If $(b_1, b_2) = (0,0)$, then we obtain $\mathcal{M}_5^6$.

Case 1.b. Suppose that L_2 is a multiple of L_1, i.e., that $L_2 = cL_1$. In this case we may express $Q(\mathcal{M}) = \mathrm{span}\{e^{cx^2}\cos(x^2), e^{cx^2}\sin(x^2), e^{L_3}\}$. By Lemma 12.33, L_3 is not independent of x^1.

   1.3. Make a linear change of coordinates to assume $L_3 = x^1$ and obtain $\mathcal{M}_5^4(c)$.

Case 2. Suppose that $Q(\mathcal{M}) = \mathrm{span}\{e^{L_1}, e^{L_2}, e^{L_3}\}$. By Lemma 12.14, $\mathrm{span}\{dL_1, dL_2, dL_3\}$ is 2-dimensional. Choose the notation so that dL_1 and dL_2 are linearly independent. Change coordinates to assume that $x^1 = L_1$ and $x^2 = L_2$. We then have

$$Q(\mathcal{M}) = \mathrm{span}\{e^{x^1}, e^{x^2}, e^{a_1 x^1 + a_2 x^2}\}.$$

If $a_1 + a_2 = 1$, then $a_1 - 1 = -a_2$ and $Q(\mathcal{M}) = e^{x^1}\mathrm{span}\{1, e^{x^2 - x^1}, e^{a_2(x^2 - x^1)}\}$. This contradicts Lemma 12.33. Thus, $a_1 + a_2 \neq 1$. A brief calculation yields:

$$\mathcal{M} = \mathcal{M}\left(\frac{a_1^2 + a_2 - 1, a_1^2 - a_1, a_1 a_2, a_1 a_2, a_2^2 - a_2, a_1 + a_2^2 - 1}{a_1 + a_2 - 1}\right),$$

$$\rho_{\mathcal{M}} = \begin{pmatrix} \frac{(a_1 - 1)a_1}{a_1 + a_2 - 1} & \frac{a_1 a_2}{a_1 + a_2 - 1} \\ \frac{a_1 a_2}{a_1 + a_2 - 1} & \frac{(a_2 - 1)a_2}{a_1 + a_2 - 1} \end{pmatrix} \quad \text{and} \quad \det(\rho_{\mathcal{M}}) = -\frac{a_1 a_2}{a_1 + a_2 - 1}.$$

   2.1. If $a_1 a_2 \neq 0$, then $\mathrm{Rank}(\rho) = 2$ and $\mathcal{M}$ is linearly equivalent to $\mathcal{M}_1^2(a_1, a_2)$.

   2.2. If $a_1 = a_2 = 0$, then $\mathrm{Rank}(\rho) = 0$ and we replace x^1 by $-x^1$ to see that $\mathcal{M}$ is linearly equivalent to $\mathcal{M}_2^6$.

   2.3. If $a_1 \neq 0$ and $a_2 = 0$ (the case $a_1 = 0$ and $a_2 \neq 0$ is analogous), then
$$Q(\mathcal{M}) = \mathrm{span}\{e^{x^1}, e^{x^2}, e^{a_1 x^1}\} \quad \text{for} \quad a_1 \neq 0, 1.$$
Change variables to replace x^1 by $c_1 x^1$ and obtain $Q(\mathcal{M}) = \mathrm{span}\{e^{c_1 x^1}, e^{x^2}, e^{a_1 c_1 x^1}\}$. Setting $a_1 = \frac{c_1 + 1}{c_1}$ then yields $Q(\mathcal{M}) = \mathrm{span}\{e^{c_1 x^1}, e^{x^2}, e^{(c_1 + 1)x^1}\}$. Since $a_1 \neq 0$, we

have that $c_1 \neq 0, -1$. Interchanging the roles of x^1 and x^2 yields

$$\mathcal{Q}(\mathcal{M}) = \text{span}\{e^{c_1 x^2}, e^{(c_1+1)x^2}, e^{x^1}\}.$$

Replacing x^1 by $c_1 x^2 - x^1$ then yields $\mathcal{M}_2^4(c_1)$.

Case 3. Suppose that $\mathcal{Q}(\mathcal{M}) = \text{span}\{e^{L_1}, L_2 e^{L_1}, e^{L_3}\}$.

Case 3.a. Suppose that $L_1 \neq 0$ and that L_1 and L_3 are linearly independent. Change coordinates to ensure $L_1 = x^1$, $L_2 = a_1 x^1 + a_2 x^2$ and $L_3 = x^1 + x^2$ so

$$\mathcal{Q}(\mathcal{M}) = e^{x^1} \text{span}\{1, a_1 x^1 + a_2 x^2, e^{x^2}\}.$$

By Lemma 12.33, $a_1 \neq 0$. We rescale $a_1 = 1$ to obtain $\mathcal{Q}(\mathcal{M}) = e^{x^1} \text{span}\{1, x^1 + a_2 x^2, e^{x^2}\}$.

3.1. If $a_2 \neq 0$, we obtain $\mathcal{M}_3^2(-a_2)$.

3.2. If $a_2 = 0$, we obtain $\mathcal{Q}(\mathcal{M}) = \text{span}\{e^{x^1}, x^1 e^{x^1}, e^{x^1+x^2}\}$. We make a linear change of coordinates to assume $\mathcal{Q}(\mathcal{M}) = \text{span}\{e^{x^2}, x^2 e^{x^2}, e^{x^2-x^1}\}$ and obtain $\mathcal{M}_1^4$.

Case 3.b. Suppose that $L_1 \neq 0$ and that $L_3 = a L_1$ for $a \neq 1$. Change coordinates to ensure that $L_1 = x^1$ and $L_2 = a_1 x^1 + a_2 x^2$ so $\mathcal{Q}(\mathcal{M}) = \text{span}\{e^{x^1}, e^{ax^1}, (a_1 x^1 + a_2 x^2)e^{x^1}\}$. By Lemma 12.33, $a_2 \neq 0$ so after replacing $a_1 x^1 + a_2 x^2$ by x^2, we obtain

$$\mathcal{Q}(\mathcal{M}) = \text{span}\{e^{x^1}, e^{ax^1}, x^2 e^{x^1}\}.$$

Replace x^1 by cx^1 to obtain $\mathcal{Q}(\mathcal{M}) = \text{span}\{e^{cx^1}, e^{acx^1}, x^2 e^{cx^1}\}$. Set $a = \frac{c+1}{c}$; we then have that $c \neq 0$. We then have $\mathcal{Q}(\mathcal{M}) = \text{span}\{e^{cx^1}, e^{(c+1)x^1}, x^2 e^{cx^1}\}$.

3.3. If $c \neq -1$, we obtain $\mathcal{M}_3^4(c)$.

3.4. If $c = -1$, then $\mathcal{Q}(\mathcal{M}) = \text{span}\{e^{-x^2}, x^1 e^{-x^2}, 1\}$ which is linearly equivalent to $\mathcal{M}_1^6$.

Case 3.c. Suppose $L_1 = 0$. Since $L_3 \neq 0$, we may change coordinates to assume $L_3 = x^2$. We then have L_2 is not a multiple of L_3, so we may assume $L_2 = x^1$.

3.5. $Q(\mathcal{M}) = \text{span}\{1, x^1, e^{x^2}\}$ and we obtain $\mathcal{M}_3^6$.

Case 4. Suppose $\mathcal{Q}(\mathcal{M}) = e^{L_1} \text{span}\{1, L_2, Q\}$. Set $\tilde{\mathcal{M}} = w^{-L_1} \mathcal{M}$; $\mathcal{Q}(\tilde{\mathcal{M}}) = \text{span}\{1, L_2, Q\}$.

Case 4.a. If $Q = L_3$ is linear, then $\mathcal{Q}(\tilde{\mathcal{M}}) = \text{span}\{1, L_2, L_3\}$. Since L_2 and L_3 are linearly independent, $\mathcal{Q}(\tilde{\mathcal{M}}) = \text{span}\{1, x^1, x^2\}$.

4.1. If $L_1 = 0$, then we obtain $\mathcal{M}_0^6$.

4.2. If $L_1 \neq 0$, choose coordinates to assume $L_1 = x^2$. We then obtain $\mathcal{M}_4^4(0)$.

Case 4.b. If Q is genuinely quadratic, change coordinates to assume $L_2 = x^2$. Since $\partial_{x^1} Q$ belongs to $\mathcal{Q}(\tilde{\mathcal{M}})$ and since $\partial_{x^1} Q$ is a multiple of x^2, $(x^1)^2$ does not appear in Q. Since $\partial_{x^2} Q$ is a multiple of x^2, $x^1 x^2$ does not appear in Q so $Q = (x^2)^2 + a_1 x^1 + a_2 x^2$. Subtract a multiple of x^2 to assume $a_2 = 0$ so $\mathcal{Q}(\tilde{\mathcal{M}}) = \text{span}\{1, x^2, (x^2)^2 + a_1 x^1\}$. Lemma 12.33 ensures $a_1 \neq 0$, so we rescale x^1 to get $\mathcal{Q}(\tilde{\mathcal{M}}) = \text{span}\{1, x^2, (x^2)^2 + 2x^1\}$.

4.3. If $L_1 = 0$, then we obtain $\mathcal{M}_4^6$.

Suppose that $L_1 \neq 0$ and $\mathcal{Q}(\mathcal{M}) = e^{b_1 x^1 + b_2 x^2}$ span$\{ \mathbb{1}, x^2, (x^2)^2 + 2x^1 \}$.

4.4. Suppose $b_1 = 0$. Set $\tilde{x}^2 := b_2 x^2$ so $\mathcal{Q}(\mathcal{M}) = e^{\tilde{x}^2}$ span$\{ \mathbb{1}, \tilde{x}^2, (2x^1 + b_2^{-2}(\tilde{x}^2)^2) \}$. Set $c = b_2^{-2} \neq 0$ to see that we have $\mathcal{M}_4^4(c)$; we obtained $\mathcal{M}_4^4(0)$ previously in 4.2.

4.5. Suppose $b_1 \neq 0$. Let $b_1 = \pm c^2$ and $\tilde{x}^2 = cx^2$ setting $\tilde{x}^1 = b_1 x^1 + b_2 x^2$. We have
$$\mathcal{Q}(\mathcal{M}) = e^{\tilde{x}^1} \text{span}\{ \mathbb{1}, x^2, (x^2)^2 + 2b_1^{-1}(\tilde{x}^1 - b_2 x^2) \}$$
$$= e^{\tilde{x}^1} \text{span}\{ \mathbb{1}, x^2, b_1(x^2)^2 + 2\tilde{x}^1 \} = e^{\tilde{x}^1} \text{span}\{ \mathbb{1}, \tilde{x}^2, \pm(\tilde{x}^2)^2 + 2\tilde{x}^1 \}.$$
We obtain $\mathcal{M}_4^2(\pm 1)$.

This proves Lemma 13.24 and thereby completes the proof of Theorem 13.22.   □

We now draw some consequences of Theorem 13.22.

13.3.2 RANK(ρ) AND DIM($\mathfrak{K}(\mathcal{M})$).

We first relate the rank of the Ricci tensor and the dimension of the space of Killing vector fields for the Type $\mathcal{A}$ geometries by extending Lemma 13.4.

Corollary 13.25   *If $\mathcal{M} = (\mathbb{R}^2, \nabla)$ is a Type $\mathcal{A}$ model, then exactly one of the following possibilities holds.*

1. *$\mathcal{M}$ is flat, $\mathrm{Rank}(\rho) = 0$, and $\dim(\mathfrak{K}(\mathcal{M})) = 6$.*
2. *$\mathrm{Rank}(\rho) = 1$ and $\dim(\mathfrak{K}(\mathcal{M})) = 4$.*
3. *$\mathrm{Rank}(\rho) = 2$ and $\dim(\mathfrak{K}(\mathcal{M})) = 2$.*

Proof. By Theorem 13.22, $\mathcal{M}$ is linearly isomorphic to one of the geometries $\mathcal{M}_i^j(\cdot)$. If $j = 6$, Assertion 1 holds. If $j = 4$, Assertion 2 holds. If $j = 2$, Assertion 3 holds.   □

13.3.3 STRONG PROJECTIVE AND LINEAR EQUIVALENCE.

We showed in Lemma 13.6 that every Type $\mathcal{A}$ affine model $\mathcal{M}$ is strongly linearly projectively equivalent to a flat Type $\mathcal{A}$ affine model $\tilde{\mathcal{M}}$. The following result now follows by inspection from the definitions given and from Theorem 12.34; it describes the extent to which $\tilde{\mathcal{M}}$ is not unique.

Theorem 13.26   *The Type $\mathcal{A}$ models $\mathcal{M}_1^6$ and $\mathcal{M}_3^6$ are strongly linearly projectively equivalent; otherwise $\mathcal{M}_i^6$ and $\mathcal{M}_j^6$ are not strongly linearly projectively equivalent for $i \neq j$.*

13.3.4 THE α INVARIANT.

By Lemma 13.3, a Type $\mathcal{A}$ model $\mathcal{M}$ is recurrent if and only if $\mathrm{Rank}(\rho) = 1$; these are the models $\mathcal{M}_i^4(\cdot)$ given in Definition 13.7. Let α be the invariant of Lemma 12.22. We use Lemma 13.8 to see:

Lemma 13.27 $\alpha\{\mathcal{M}_1^4\} = 16$, $\alpha\{\mathcal{M}_2^4(c_1)\} = 4\frac{(1+2c_1)^2}{c_1 + c_1^2}$, $\alpha\{\mathcal{M}_3^4(c_1)\} = 4\frac{(1+2c_1)^2}{c_1 + c_1^2}$,

$\alpha\{\mathcal{M}_4^4(c)\} = 16$, $\alpha\{\mathcal{M}_5^4(c)\} = \frac{16c^2}{1+c^2}$.

This invariant together with the sign of ρ is a complete invariant in this setting.

Theorem 13.28   *Let $\mathcal{M}_1$ and $\mathcal{M}_2$ be two Type $\mathcal{A}$ models whose Ricci tensor has rank 1. Assume that $\alpha(\mathcal{M}_1) = \alpha(\mathcal{M}_2)$. If $\alpha(\mathcal{M}_1) = 0$, also assume either $\rho_{\mathcal{M}_1} \geq 0$ and $\rho_{\mathcal{M}_2} \geq 0$ or that $\rho_{\mathcal{M}_1} \leq 0$ and $\rho_{\mathcal{M}_2} \leq 0$. Then $\mathcal{M}_1$ and $\mathcal{M}_2$ are locally isomorphic.*

Proof. By Theorem 13.22, it suffices to prove this result where the models in question belong to the families $\mathcal{M}_i^4(\cdot)$. By Lemma 13.12, $\mathcal{M}_1^4$ is locally affine isomorphic to $\mathcal{M}_4^4(0)$, $\mathcal{M}_4^4(c)$ is locally affine isomorphic to $\mathcal{M}_4^4(0)$, and $\mathcal{M}_2^4(c_1)$ is locally affine isomorphic to $\mathcal{M}_3^4(c_1)$. Thus we must only show that these invariants distinguish the spaces $\mathcal{M}_3^4(c_1)$, $\mathcal{M}_4^4(0)$, and $\mathcal{M}_5^4(c)$. We have $\mathcal{M}_3^4(c_1)$ is symmetric around the axis $c_1 = -\frac{1}{2}$ and $\mathcal{M}_5^4(c)$ is symmetric around the axis $c = 0$. By Lemma 13.27,

$$\alpha\{\mathcal{M}_3^4(c_1)\} = 4\frac{(1+2c_1)^2}{c_1+c_1^2}, \quad \alpha\{\mathcal{M}_4^4(0)\} = 16, \quad \alpha\{\mathcal{M}_5^4(c)\} = \frac{16c^2}{1+c^2}.$$

We graph below these functions. The graph of $\alpha(\mathcal{M}_3^4(t))$ is given in red; it is the upper and the lower curve and has a vertical asymptote at $t \in \{-1, 0\}$; it is symmetric around the line $t = -\frac{1}{2}$. The graph of $\alpha(\mathcal{M}_4^4(0))$ is given in brown; it takes the constant value 16. The graph of $\alpha(\mathcal{M}_5^4(t))$ is given in blue. It lies in the middle and is symmetric around the line $t = 0$. The vertical scale is compressed by a factor of $\frac{1}{16}$.

Figure 13.1: The graph of α

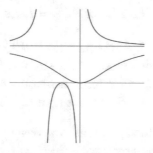

The α invariant of these three families is distinct except for the values

$$\alpha\{\mathcal{M}_5^4(0)\} = 0 \quad \text{and} \quad \alpha\{\mathcal{M}_3^4(-\tfrac{1}{2})\} = 0;$$

these two manifolds are affine symmetric spaces. We have $\rho_{\mathcal{M}_3^4(-\frac{1}{2})} \leq 0$ and $\rho_{\mathcal{M}_5^4(0)} \geq 0$ so the sign of the Ricci tensor distinguishes these two spaces. The function $4\frac{(1+2t)^2}{t+t^2}$ is 1-1 on $(-\infty, -\frac{1}{2}) \cup (-\frac{1}{2}, 0)$ and the function $\frac{16t^2}{1+t^2}$ is 1-1 on $(-\infty, 0)$. $\qquad\square$

The affine symmetric spaces $\mathcal{M}_5^4(0)$ and $\mathcal{M}_3^4(-\frac{1}{2})$ were already considered by Opozda [50], who provided a parameterization of $\tilde{\mathcal{M}}_5^4(0)$.

13.3.5 GEODESIC COMPLETENESS. Let $\mathcal{M}$ be a Type $\mathcal{A}$ model. Let $\sigma_{a,b}(t)$ be the affine geodesic in $\mathcal{M}$ so that $\sigma_{a,b}(0) = (0,0)$ and $\dot{\sigma}_{a,b}(0) = (a,b)$. Our investigation was informed by Theorem 12.35 and Lemma 13.9; it was not a routine computation. When the Ricci tensor had rank 2, in the generic case, we could not in general solve the resulting ODE for the reparameterizing function and we contented ourselves with obtaining enough information for later use. We shall follow the treatment in Gilkey, Park, and Valle-Regueiro [34] and refer to D'Ascanio, Gilkey, and Pisani [21] for a different approach. We first deal with the flat geometries:

Lemma 13.29 *Let $\mathcal{M}$ be a Type $\mathcal{A}$ model with $\dim(\mathfrak{K}(\mathcal{M})) = 6$ and let $\sigma(a,b)(t)$ be an affine geodesic in $\mathcal{M}$ with $\sigma(a,b)(0) = (0,0)$ and $\dot{\sigma}(a,b)(0) = (a,b)$. Then:*

1. *Let $\mathcal{M} = \mathcal{M}_0^6$. Then $\sigma(a,b)(t) = t(a,b)$. $\mathcal{M}_0^6$ is geodesically complete.*
2. *Let $\mathcal{M} = \mathcal{M}_1^6$. Then $\sigma(a,b)(t) = (\log(1 + at), (1 + at)^{-1}bt)$. $\mathcal{M}_1^6$ is geodesically incomplete.*
3. *Let $\mathcal{M} = \mathcal{M}_2^6$. Then $\sigma(a,b)(t) = (-\log(1 - at), \log(1 + bt))$. $\mathcal{M}_2^6$ is geodesically incomplete.*
4. *Let $\mathcal{M} = \mathcal{M}_3^6$. Then $\sigma(a,b)(t) = (at, \log(1 + bt))$. $\mathcal{M}_3^6$ is geodesically incomplete.*
5. *Let $\mathcal{M} = \mathcal{M}_4^6$. Then $\sigma(a,b)(t) = (at \quad \frac{1}{2}b^2 t^2, bt)$. $\mathcal{M}_4^6$ is geodesically complete.*
6. *Let $\mathcal{M} = \mathcal{M}_5^6$. Then*

$$\sigma(a,b)(t) = \left(\tfrac{1}{2}\log(1 + 2at + a^2 t^2 + b^2 t^2), \arctan\left(\tfrac{1+(a+b)t}{1+at-bt} \right) - \tfrac{\pi}{4} \right).$$

$\mathcal{M}_5^6$ is geodesically incomplete.

Proof. A direct computation verifies that the curves given satisfy the geodesic equation for the structures given with initial position $\sigma(0) = (0,0)$ and initial velocity $\dot{\sigma}(0) = (a,b)$. The curves are defined for all time if $\mathcal{M} = \mathcal{M}_0^6$ or $\mathcal{M} = \mathcal{M}_4^6$. This shows that these structures are geodesically complete. In the remaining structures, some of the affine geodesics blow up at finite time for some value of the parameters so these structures are geodesically incomplete. By Lemma 13.12, all these structures have an affine embedding or immersion in $\mathcal{M}_0^6$, so they can all be geodesically completed. □

We now turn to the geometries where the Ricci tensor has rank 1. By Lemma 12.8, if there exists an affine geodesic so $\lim_{t \to T} \rho(\dot{\sigma}(t), \dot{\sigma}(t)) = \pm\infty$, the geometry is essentially geodesically incomplete. In all these geometries, $\rho = \lambda\, dx^2 \otimes dx^2$ where $\lambda \neq 0$. Thus we shall examine $\dot{x}^2$ with particular care. Since we can solve the geodesic equations explicitly, the following result is now immediate.

Lemma 13.30 *Let $\mathcal{M}$ be a Type $\mathcal{A}$ model with $\dim(\mathfrak{K}(\mathcal{M})) = 4$ and let $\sigma(a,b)(t)$ be an affine geodesic in $\mathcal{M}$ with $\sigma(a,b)(0) = (0,0)$ and $\dot{\sigma}(a,b)(0) = (a,b)$. Then:*

1. Let $\mathcal{M} = \mathcal{M}_1^4$.

(a) *If* $b \neq 0$, *then* $\sigma(a,b)(t) = \left(-\log\left(1 - \frac{a\log(1+2bt)}{2b}\right), \frac{1}{2}\log(1 + 2bt)\right)$.

(b) *If* $b = 0$, *then* $\sigma(a,0)(t) = (-\log(1-at), 0)$.

(c) *$\mathcal{M}$ is essentially geodesically incomplete.*

2.1. Let $\mathcal{M} = \mathcal{M}_2^4(c_1)$, $c_1 \neq -\frac{1}{2}$.

(a) *If* $b \neq 0$ *and* $a + b \neq 0$, *then*

$$\sigma(a,b)(t) = \left(\log(\tfrac{b}{a+b}) - \log(1 - \tfrac{a(2bc_1t+bt+1)^{\frac{1}{2c_1+1}}}{a+b}), \frac{\log(2bc_1t+bt+1)}{2c_1+1}\right).$$

(b) *If* $b \neq 0$ *and* $a + b = 0$, *then* $\sigma(-b,b)(t) = \left(-\frac{\log(1+bt+2bc_1t)}{1+2c_1}, \frac{\log(1+bt+2bc_1t)}{1+2c_1}\right)$.

(c) *If* $b = 0$, *then* $\sigma(a,0)(t) = (-\log(1-at), 0)$.

(d) *$\mathcal{M}$ is essentially geodesically incomplete.*

2.2. Let $\mathcal{M} = \mathcal{M}_2^4(-\frac{1}{2})$.

(a) *If* $b \neq 0$ *and* $a \neq 0$, *then* $\sigma(a,b)(t) = \left(-\log\left(\frac{a+b-ae^{bt}}{b}\right), bt\right)$.

(b) *If* $b \neq 0$ *and* $a = 0$, *then* $\sigma(0,b)(t) = (0, bt)$.

(c) *If* $b = 0$ *and* $a \neq 0$, *then* $\sigma(a,0)(t) = (-\log(1-at), 0)$.

(d) *$\mathcal{M}$ is geodesically incomplete. However, $\mathcal{M}$ affine embeds in $\mathcal{M}_3^4(-\frac{1}{2})$ which is geodesically complete.*

3.1. Let $\mathcal{M} = \mathcal{M}_3^4(c_1)$, $c_1 \neq -\frac{1}{2}$.

(a) *If* $b \neq 0$, *then*

$$\sigma(a,b)(t) = \left(\tfrac{a}{b}(-1 + (1 + b(t + 2c_1t))^{\frac{1}{1+2c_1}}), \tfrac{1}{1+2c_1}\log(1 + bt + 2bc_1t)\right).$$

(b) *If* $b = 0$, *then* $\sigma(a,0)(t) = (at, 0)$.

(c) *$\mathcal{M}$ is essentially geodesically incomplete.*

3.2. Let $\mathcal{M} = \mathcal{M}_3^4(-\frac{1}{2})$.

(a) *If* $b \neq 0$, *then* $\sigma(a,b)(t) = \left(\frac{a}{b}(e^{bt} - 1), bt\right)$.

(b) *If* $b = 0$, *then* $\sigma(a,0)(t) = (at, 0)$.

(c) *$\mathcal{M}$ is geodesically complete.*

4. Let $\mathcal{M} = \mathcal{M}_4^4(c)$ and $\tau(t) := 1 + 2bt$.

(a) *If* $b \neq 0$, *then* $\sigma(a,b)(t) = \left(-\frac{\log(\tau(t))(bc\log(\tau(t))-4a)}{8b}, \frac{1}{2}\log(\tau(t))\right)$.

(b) *If* $b = 0$, *then* $\sigma(a,0)(t) = (at, 0)$.

(c) *$\mathcal{M}$ is essentially geodesically incomplete.*

5.1. Let $\mathcal{M} = \mathcal{M}_5^4(c)$, $c \neq 0$ and $\tau(t) := \log(1 + 2bct)$.

(a) If $b \neq 0$, then $\sigma(a,b)(t) = \left(\log\left(\frac{(a-bc)\sin\left(\frac{\tau(t)}{2c}\right)}{b} + \cos\left(\frac{\tau(t)}{2c}\right) \right) + \frac{1}{2}\tau(t), \frac{\tau(t)}{2c} \right).$

(b) If $b = 0$, then $\sigma(a,0)(t) = (\log(1 + at), 0).$

(c) $\mathcal{M}$ is essentially geodesically incomplete.

5.2. Let $\mathcal{M} = \mathcal{M}_5^4(0).$

  (a) If $b \neq 0$, then $\sigma(a,b)(t) = \left(\frac{1}{2}\log\left(\frac{a^2}{b^2} + 1 \right) + \log\left(\cos\left(bt - \arctan\left(\frac{a}{b}\right) \right) \right), bt \right).$

  (b) If $b = 0$, then $\sigma(a,0)(t) = (\log(1 + at), 0).$

  (c) $\mathcal{M}$ is geodesically incomplete. However, $\mathcal{M}$ affine embeds in the geodesically complete manifold $\tilde{\mathcal{M}}_5^4(0).$

5.3. Let $\mathcal{M} = \tilde{\mathcal{M}}_5^4(c)$, $c \neq 0.$

  (a) If $b \neq 0$, then $\sigma(a,b)(t) = \left(\frac{a}{b}(1 + 2bct)^{1/2} \sin(\frac{\log(1+2bct)}{2c}), \frac{\log(1+2bct)}{2c} \right).$

  (b) If $b = 0$, then $\sigma(a,0)(t) = (at, 0).$

  (c) $\mathcal{M}$ is essentially geodesically incomplete.

5.4. Let $\mathcal{M} = \tilde{\mathcal{M}}_5^4(0).$

  (a) If $b \neq 0$, then $\sigma(a,b)(t) = \left(\frac{a}{b}\sin(bt), bt \right).$

  (b) If $b = 0$, then $\sigma(a,0)(t) = (at, 0).$

  (c) $\mathcal{M}$ is geodesically complete

Let $\mathcal{M}$ be a Type $\mathcal{A}$ model with $\mathrm{Rank}(\rho) = 2$. By Lemma 12.8, if there exists an affine geodesic and an affine Killing vector field X so $\lim_{t \to T} \rho(\dot{\sigma}(t), X(\sigma(t))) = \pm\infty$ for some finite time T, then $\mathcal{M}$ is essentially geodesically incomplete. Since ∂_{x^1} and ∂_{x^2} are affine Killing vector fields and since ρ is non-degenerate, $\mathcal{M}$ is essentially geodesically incomplete if

$$\lim_{t \to T} \dot{x}^1(t) = \pm\infty \text{ or if } \lim_{t \to T} \dot{x}^2(t) = \pm\infty.$$

We were unable to solve the ODE for the geodesic equation in closed form for the geometries $\mathcal{M}_i^2$ where the Ricci tensor has rank 2. However, we were able to obtain sufficient information to establish the following result.

Lemma 13.31 *Let $\mathcal{M}$ be a Type $\mathcal{A}$ model with $\dim(\mathfrak{K}(\mathcal{M})) = 2$ and let $\sigma(a,b)(t)$ be an affine geodesic in $\mathcal{M}$ with $\sigma(a,b)(0) = (0,0)$ and $\dot{\sigma}(a,b)(0) = (a,b)$. Then:*

1. *$\mathcal{M}_1^2(a_1, a_2)$ is essentially geodesically incomplete.*
2. *$\mathcal{M}_3^2(c_2)$ and $\mathcal{M}_4^2(\pm 1)$ are essentially geodesically incomplete.*
3. *If $b_1 \neq -1$, then $\mathcal{M}_2^2(b_1, b_2)$ is essentially geodesically incomplete.*
4. *$\mathcal{M}_2^2(-1, b_2)$ is geodesically complete.*

Proof. We apply the criteria of Lemma 12.8 to prove the first three assertions; the proof of the final assertion requires a more delicate argument.

Assertion 1. $\mathcal{M} = \mathcal{M}_1^2(a_1, a_2)$. We obtain three possible affine geodesics which have the form $\sigma_i(t) = \log(t)\vec{\alpha}_i$ where

$$\vec{\alpha}_1 = \tfrac{1}{1+a_1+a_2}(1,1), \quad \vec{\alpha}_2 = \tfrac{1}{1+a_1-a_2}(1-a_2, a_1), \quad \vec{\alpha}_3 = \tfrac{1}{1-a_1+a_2}(a_2, 1-a_1).$$

The first affine geodesic is defined for $a_1 + a_2 + 1 \neq 0$, the second for $a_1 - a_2 + 1 \neq 0$, and the third for $-a_1 + a_2 + 1 \neq 0$. At least two affine geodesics are defined for any given geometry. These geometries are essentially geodesically incomplete.

Assertion 2. $\mathcal{M} = \mathcal{M}_3^2(c_2)$ or $\mathcal{M} = \mathcal{M}_4^2(\pm 1)$. These geometries are essentially geodesically incomplete since we have an affine geodesic of the form $\sigma(t) = (\tfrac{1}{2}\log(1+t), 0)$.

Assertion 3. $\mathcal{M} = \mathcal{M}_2^2(b_1, b_2)$ for $b_1 \neq -1$. These geometries are essentially geodesically incomplete since we have an affine geodesic of the form $\sigma(t) = \log(t)(\tfrac{1}{1+b_1}, 0)$.

Assertion 4. $\mathcal{M} = \mathcal{M}_2^2(-1, b_2)$. Suppose, to the contrary, that $\mathcal{M}$ is geodesically incomplete. Let σ be an affine geodesic in $\mathcal{M}$ which is defined on a parameter range (t_0, t_1) where $t_1 < \infty$ (resp. $-\infty < t_0$) which cannot be extended to a parameter range $(t_0, t_1 + \varepsilon)$ (resp. $(t_0 - \varepsilon, t_1)$) for any $\varepsilon > 0$. By Lemma 12.7, this implies that $\lim_{t \downarrow t_0^+} \sigma(t)$ (resp. $\lim_{t \uparrow t_1^-} \sigma(t)$) does not exist. We argue for a contradiction; our proof is motivated by the paper of Bromberg and Medina [4]. The non-zero Christoffel symbols of $\mathcal{M}$ are $\Gamma_{12}{}^1 = b_2$, $\Gamma_{12}{}^2 = 1$, and $\Gamma_{22}{}^1 = -\tfrac{1}{2}(1 + b_2^2)$. We work in the tangent bundle and introduce variables $u^1(t) := \dot{x}^1(t)$ and $u^2(t) := \dot{x}^2(t)$. This yields the geodesic equations

$$\dot{u}^1 + 2b_2 u^1 u^2 - \tfrac{1}{2}(1 + b_2^2)u^2 u^2 = 0 \quad \text{and} \quad \dot{u}^2 + 2u^1 u^2 = 0. \tag{13.3.a}$$

If $u^2(t) = 0$ for any $s \in (t_0, t_1)$, then $\dot{u}^1(s) = 0$ and $\dot{u}^2(s) = 0$. Consequently, $u^1(t) = u^1(s)$ and $u^2(t) = u^2(s)$ solves this ODE and $\vec{u} = (u^1, u^2)$ is constant on the interval (t_0, t_1). We may therefore assume u^2 does not change on the interval (t_0, t_1). Set

$$u^1(t) := e^{-b_2\tau(t)}\left(\tfrac{1}{2}\left(-2ab_2 + bb_2^2 + b\right)\sin(\tau(t)) + a\cos(\tau(t))\right),$$
$$u^2(t) := e^{-b_2\tau(t)}((bb_2 - 2a)\sin(\tau(t)) + b\cos(\tau(t)))$$

where τ is a function which needs to be determined. Equation (13.3.a) then gives rise to a single ODE to be satisfied: $(2a - bb_2)\sin(\tau(t)) - b\cos(\tau(t)) + e^{b_2\tau(t)}\tau'(t)$, i.e., we have $\dot{\tau}(t) = u^2(\tau(t))$. Since u^2 does not change sign, $\tau(t)$ is restricted to a parameter interval of length at most π. If u^2 is positive (resp. negative), then $\dot{\tau}(t)$ is positive (resp. negative) and bounded so $\tau(t)$ is monotonically increasing (resp. decreasing) and bounded on the interval (t_0, t_1). Thus $\lim_{t \downarrow t_0} \tau(t)$ and $\lim_{t \uparrow t_1} \tau(t)$ exist. This shows that $\lim_{t \downarrow t_0} \dot{\sigma}(t)$ and $\lim_{t \uparrow t_1} \dot{\sigma}(t)$ exist. We integrate to conclude $\lim_{t \downarrow t_0} \dot{\sigma}(t)$ and $\lim_{t \uparrow t_1} \dot{\sigma}(t)$ exist. This provides the desired contradiction. $\square$

The next result follows from the previous analysis and from Theorem 13.22.

Theorem 13.32   *Let $\mathcal{M}$ be a Type $\mathcal{A}$ model.*

1. *If the Ricci tensor of $\mathcal{M}$ is zero, then $\mathcal{M}$ is geodesically complete if and only if $\mathcal{M}$ is linearly equivalent to $\mathcal{M}_0^6$ or to $\mathcal{M}_4^6$.*

2. *If the Ricci tensor of $\mathcal{M}$ has rank 1, then $\mathcal{M}$ is geodesically complete if and only if $\mathcal{M}$ is linearly equivalent to $\mathcal{M}_3^4(-\frac{1}{2})$.*

3. *If the Ricci tensor of $\mathcal{M}$ has rank 2, then $\mathcal{M}$ is geodesically complete if and only if $\mathcal{M}$ is linearly equivalent to $\mathcal{M}_2^2(-1, b)$ for some b.*

4. *If $\tilde{\mathcal{M}}$ is a geodesically complete homogeneous surface which is modeled on $\mathcal{M}$, then $\mathcal{M}$ can be taken to be $\mathcal{M}_0^6$, $\mathcal{M}_3^4(-\frac{1}{2})$, $\mathcal{M}_5^4(0)$, or $\mathcal{M}_2^2(-1, b)$ for some b.*

5. *The models $\{\mathcal{M}_i^6, \mathcal{M}_2^4(-\frac{1}{2}), \mathcal{M}_5^4(0)\}$, for $i \notin \{0, 4\}$, are geodesically incomplete. These models can be geodesically completed. All the remaining models $\mathcal{M}_i^j(\cdot)$ which are geodesically incomplete are essentially geodesically incomplete.*

We give below pictures of the geodesic structures for the models $\mathcal{M}_0^6$, $\mathcal{M}_3^4(-\frac{1}{2})$, $\tilde{\mathcal{M}}_5^4(0)$, and $\mathcal{M}_2^2(-1, b)$ for $b = 0, 1, 2$. The pictures for $\mathcal{M}_0^6$, $\mathcal{M}_3^4(-\frac{1}{2})$, and $\tilde{\mathcal{M}}_5^4(0)$ arise from the parameterizations obtained in this section; the pictures for $\mathcal{M}_2^2(\cdot)$ were obtained numerically.

Figure 13.2: Geodesic structure of $\mathcal{M}_0^6$, $\mathcal{M}_4^6$, $\mathcal{M}_3^4(-\frac{1}{2})$, $\tilde{\mathcal{M}}_5^4(0)$

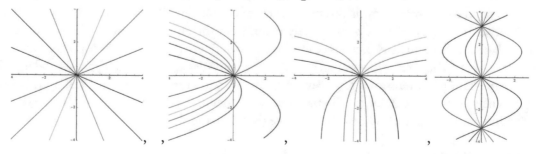

Figure 13.3: Geodesics for $\mathcal{M}_2^2(-1,0)$, $\mathcal{M}_2^2(-1,1)$, $\mathcal{M}_2^2(-1,2)$

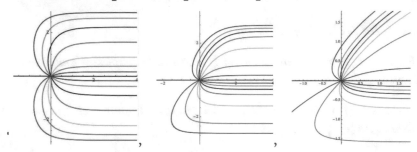

13.4   TYPE $\mathcal{A}$: MODULI SPACES

If $\mathcal{M}$ is a Type $\mathcal{A}$ model, let $\mathcal{G}(\mathcal{M}) := \{T \in \mathrm{GL}(2,\mathbb{R}) : T^*\mathcal{M} = \mathcal{M}\}$ be the associated symmetry group; by Lemma 12.39,

$$\mathcal{G}(\mathcal{M}) = \{T \in \mathrm{GL}(2,\mathbb{R}) : T^*\mathcal{Q}(\mathcal{M}) = \mathcal{Q}(\mathcal{M})\}.$$

Let $\mathcal{S}(\mathcal{M})$ be the set of all Type $\mathcal{A}$ models which are linearly equivalent to $\mathcal{M}$. Since $\mathrm{GL}(2,\mathbb{R})$ acts smoothly and transitively on $\mathcal{S}(\mathcal{M})$, $\mathcal{S}(\mathcal{M})$ is a homogeneous space and

$$\mathcal{S}(\mathcal{M}) = \mathrm{GL}(2,\mathbb{R})/\mathcal{G}(\mathcal{M}).$$

In Section 13.4.1, we study flat Type $\mathcal{A}$ connections. In Theorem 13.33, we determine the symmetry groups $\mathcal{G}(\mathcal{M}_i^6)$ and thereby determine the spaces $\mathcal{S}(\mathcal{M}_i^6)$. Let

$$\mathcal{S}_{\mathcal{A}}^6 = \cup_{1 \leq i \leq 5} \mathcal{S}(\mathcal{M}_i^6)$$

be the space of all flat Type $\mathcal{A}$ models other than the trivial model $\mathcal{M}_0^6$ which otherwise would be a cone point. Let $\mathbb{L}$ be the Möbius strip over the circle and let $⫫$ be the trivial real line bundle over the circle. In Theorem 13.34, we show that $\mathcal{S}_{\mathcal{A}}^6$ is diffeomorphic to $\mathbb{L} \oplus ⫫ \oplus ⫫$ minus the zero section. In Section 13.4.2 (see Theorem 13.35), we perform a similar analysis for the set of all $\mathcal{A}$ models with rank 1 Ricci tensor; this space naturally decomposes into two components depending on whether ρ is positive or negative semi-definite. Note that the α invariant together with the signature of ρ completely determines the affine equivalence classes by Theorem 13.28. In Section 13.4.3, we discuss the two invariants (ψ, Ψ) which, together with the signature of the Ricci tensor, completely detect the equivalence class of a Type $\mathcal{A}$ model with non-degenerate Ricci tensor up to linear equivalence or, in view of Theorem 13.5, up to affine equivalence. In Section 13.4.4, we discuss linear equivalence for the manifolds $\mathcal{M}_1^2(a_1, a_2)$, and in Section 13.4.5, we discuss linear equivalence for the manifolds $\mathcal{M}_2^2(b_1, b_2)$.

13.4.1 FLAT TYPE $\mathcal{A}$ CONNECTIONS. The following result is an immediate application of Theorem 12.34 and Lemma 13.9.

Theorem 13.33

1. $\mathcal{G}(\mathcal{M}_0^6) = \mathrm{GL}(2, \mathbb{R})$.

2. $\mathcal{G}(\mathcal{M}_1^6) = \left\{ \begin{pmatrix} 1 & 0 \\ 0 & a \end{pmatrix} : a \neq 0 \right\}$.

3. $\mathcal{G}(\mathcal{M}_2^6) = \left\{ \begin{pmatrix} 1 & 0 \\ 0 & 1 \end{pmatrix}, \begin{pmatrix} 0 & -1 \\ -1 & 0 \end{pmatrix} \right\}$.

4. $\mathcal{G}(\mathcal{M}_3^6) = \left\{ \begin{pmatrix} a & 0 \\ 0 & 1 \end{pmatrix} : a \neq 0 \right\}$.

5. $\mathcal{G}(\mathcal{M}_4^6) = \left\{ \begin{pmatrix} a^2 & b \\ 0 & a \end{pmatrix} : a \neq 0,\, b \in \mathbb{R} \right\}$.

6. $\mathcal{G}(\mathcal{M}_5^6) = \left\{ \begin{pmatrix} 1 & 0 \\ 0 & \pm 1 \end{pmatrix} \right\}$.

Theorem 13.34 $\mathcal{S}_{\mathcal{A}}^6$ *is a smooth submanifold of* $\mathbb{R}^6 - \{0\}$ *which is diffeomorphic to the total space of the bundle* $\mathbb{L} \oplus \mathbb{1} \oplus \mathbb{1}$ *over the circle minus the zero section.*

Proof. Let (Ξ_1, Ξ_2) be a point of the unit circle S^1 in $\mathbb{R}^2$ and let (x^1, x^2, x^3) denote a point of $\mathbb{R}^3 - (0, 0, 0)$. We identify $((\Xi_1, \Xi_2), (x^1, x^2, x^3))$ with $((-\Xi_1, -\Xi_2), (-x^1, x^2, x^3))$ to define the bundle $\mathbb{L} \oplus \mathbb{1} \oplus \mathbb{1}$. Let $\mathcal{M}(\xi_1, \xi_2, \xi_3, \xi_4, \xi_5, \xi_6)$ be the Type $\mathcal{A}$ model given in Equation (13.2.a). We make a linear change of variables and set

$$\mathcal{M}_1(p, q, t, s, v, w) := \mathcal{M}(2q, p + t, w, q + s, v, p - t).$$

We have:

$$\rho_{11} = p^2 + q^2 - s^2 - t^2 - pw - tw, \quad \rho_{12} = \rho_{21} = -(p+t)v + (q+s)w,$$
$$\rho_{22} = qv - sv + (p - t - w)w. \tag{13.4.a}$$

The equations $\rho_{12} = 0$ and $\rho_{22} = 0$ can be written in matrix form

$$\begin{pmatrix} -v & w \\ w & v \end{pmatrix} \begin{pmatrix} p \\ q \end{pmatrix} = \begin{pmatrix} tv - sw \\ sv + (t+w)w \end{pmatrix}.$$

If $v^2 + w^2 \neq 0$, we solve these equations to express

$$p = \frac{2svw + t\left(w^2 - v^2\right) + w^3}{v^2 + w^2} \quad \text{and} \quad q = \frac{s\left(v^2 - w^2\right) + vw(2t + w)}{v^2 + w^2}. \tag{13.4.b}$$

Substituting these values of p and q into Equation (13.4.a) then yields $\rho_{11} = 0$ as well; somewhat surprisingly, perhaps, there is no additional relation. This parameterizes the portion of $\mathcal{S}_{\mathcal{A}}^6$

where $(v, w) \neq (0, 0)$ as a graph over the (s, t, v, w) coordinate 4-plane. If we set $v = w = 0$, the equation $\rho = 0$ becomes the single equation

$$p^2 + q^2 - s^2 - t^2 = 0. \qquad (13.4.c)$$

There is an apparent singularity in Equation (13.4.b) at $(v, w) = (0, 0)$. We introduce polar coordinates $v = r\cos(\theta)$ and $w = r\sin(\theta)$ to obtain

$$p = p(\theta, r, s, t) := r\sin^3(\theta) + s\sin(2\theta) - t\cos(2\theta),$$
$$q = q(\theta, r, s, t) := \cos(\theta)\sin^2(\theta) + s\cos(2\theta) + t\sin(2\theta).$$

This no longer has a singularity at $r = 0$; when $r = 0$, we obtain

$$\begin{pmatrix} \sin(2\theta) & -\cos(2\theta) \\ \cos(2\theta) & \sin(2\theta) \end{pmatrix} \begin{pmatrix} s \\ t \end{pmatrix} = \begin{pmatrix} p \\ q \end{pmatrix}$$

which parameterizes the singular locus given by Equation (13.4.c). To avoid the trivial connection $\Gamma_0 = \Gamma(0, 0, 0, 0, 0, 0)$, we need to assume $(r, s, t) \neq (0, 0, 0)$; the parameterization is singular at this point since θ no longer plays a role. Thus our parameters take the form $\theta \in \mathbb{R}/(2\pi\mathbb{Z})$ and $(r, s, t) \in \mathbb{R}^3 - \{0\}$; since we are permitting r to be negative in polar coordinates, we must identify (θ, r) with $(\theta + \pi, -r)$. These are exactly the identifications defining the total space of the bundle $\mathbb{L} \oplus \mathbb{1} \oplus \mathbb{1}$ minus the zero section once we identify $\mathbb{R}/(2\pi\mathbb{Z})$ with S^1.   □

13.4.2 TYPE $\mathcal{A}$ CONNECTIONS WITH RANK 1 RICCI TENSOR.

The set of all Type $\mathcal{A}$ connections where the Ricci tensor has rank 1 is the disjoint union $\mathcal{S}^4_+ \dot\cup \mathcal{S}^4_-$ of two spaces where

$$\mathcal{S}^4_+ := \{\Gamma \in \mathbb{R}^6 : \mathrm{Rank}(\rho_\Gamma) = 1 \text{ and } \rho_\Gamma \geq 0\},$$
$$\mathcal{S}^4_- := \{\Gamma \in \mathbb{R}^6 : \mathrm{Rank}(\rho_\Gamma) = 1 \text{ and } \rho_\Gamma \leq 0\}.$$

Theorem 13.34 generalizes to this situation to become the following result.

Theorem 13.35 $\mathcal{S}^4_\pm$ *is a smooth submanifold of* $\mathbb{R}^6 - \{0\}$ *diffeomorphic to* $S^1 \times S^1 \times \mathbb{R}^3$.

Proof. Let $\mathcal{S}^4_{+,0}$ (resp. $\mathcal{S}^4_{-,0}$) be the space of all Type $\mathcal{A}$ models where the Ricci tensor is a positive (resp. negative) multiple of $dx^2 \otimes dx^2$. By Lemma 13.3, an element of $\mathcal{S}^4_{\pm,0}$ satisfies the conditions

$$\Gamma_{11}{}^2 = 0, \quad \Gamma_{12}{}^2 = 0, \quad \rho_{22} = \Gamma_{22}{}^2\Gamma_{12}{}^1 - (\Gamma_{12}{}^1)^2 + \Gamma_{11}{}^1\Gamma_{22}{}^1 \neq 0.$$

We make a change of variables setting

$$\Gamma_{11}{}^1(p, q, u, v) := q + v, \quad \Gamma_{11}{}^2(p, q, u, v) := 0, \quad \Gamma_{12}{}^1(p, q, u, v) := u + p,$$
$$\Gamma_{12}{}^2(p, q, u, v) := 0, \quad\quad \Gamma_{22}{}^1 := q - v, \quad\quad \Gamma_{22}{}^2 := 2p.$$

We then have $\rho_{22} = (p^2 + q^2 - u^2 - v^2)dx^2 \otimes dx^2$. Thus we have

$$\mathcal{S}^4_{+,0} = \{\Gamma(p,q,u,v) : p^2 + q^2 > u^2 + v^2\},$$
$$\mathcal{S}^4_{-,0} = \{\Gamma(p,q,u,v) : p^2 + q^2 < u^2 + v^2\}.$$

We examine $\mathcal{S}^4_{+,0}$ as the analysis of $\mathcal{S}^4_{-,0}$ is the same after interchanging the roles of (p,q) and (u,v). Let $\dot{D}^2 := \{(\tilde{u}, \tilde{v}) \in \mathbb{R}^2 : \tilde{u}^2 + \tilde{v}^2 < 1\}$. We construct a diffeomorphism

$$\tilde{\Gamma}(\theta, r, \tilde{u}, \tilde{v}) := \Gamma(p(\cdot), q(\cdot), u(\cdot), v(\cdot)) : S^1 \times \mathbb{R}^+ \times \dot{D}^2 \to \mathcal{S}^4_{+,0}$$

by setting $p = r\cos\theta$, $q = r\sin\theta$, $u = r\tilde{u}$, $v = r\tilde{v}$. We then have

$$-\Gamma_+(\theta, r, \tilde{u}, \tilde{v}) = \Gamma_+(\theta + \pi, r, -\tilde{u}, -\tilde{v}).$$

Consider the coordinate rotation through an angle of ϕ

$$T_\phi(x^1, x^2) = (\cos(\phi)x^1 + \sin(\phi)x^2, -\sin(\phi)x^1 + \cos(\phi)x^2).$$

Let $\tilde{\mathcal{M}}$ be an arbitrary Type $\mathcal{A}$ model with $\mathrm{Rank}(\rho_{\tilde{\mathcal{M}}}) = 1$ and $\rho_{\tilde{\mathcal{M}}}$ positive semi-definite. There exists $\lambda > 0$ and ϕ so that

$$\rho_{\tilde{\mathcal{M}}} = \lambda(-\sin(\phi)dx^1 + \cos(\phi)dx^2) \otimes (-\sin(\phi)dx^1 + \cos(\phi)dx^2);$$

the angle ϕ is not uniquely determined but only defined modulo π and not the more usual 2π. Let $\mathcal{M} := (T_\phi)_* \tilde{\mathcal{M}}$. Then $\rho_{\mathcal{M}} = \lambda dx^2 \otimes dx^2$ and, consequently, $\mathcal{M} \in \mathcal{S}^4_{+,0}$. Let $-\mathcal{M}$ be defined by $-\Gamma$. If we replace ϕ by $\phi + \pi$, then we replace $\mathcal{M}$ by $-\mathcal{M}$ and thus we have that

$$\mathcal{S}^4_+ = \{(\mathbb{R}/(2\pi\mathbb{Z})) \times \mathcal{S}^4_{+,0}\}/(\phi, \mathcal{M}) \sim (\phi + \pi, -\mathcal{M})$$

where the gluing reflects the fact that when $\phi = \pi$ we have replaced (x^1, x^2) by $(-x^1, -x^2)$ and thus replaced Γ by $-\Gamma$. Using our previous parameterization of $\mathcal{S}^4_{+,0}$, this yields

$$\mathcal{S}^4_+ = (\mathbb{R}^2/(2\pi\mathbb{Z})^2) \times \mathbb{R}^+ \times \dot{D}^2/\{(\phi, \theta, r, \tilde{u}, \tilde{v}) \sim (\phi + \pi, \theta + \pi, r, -\tilde{u}, -\tilde{v})\}.$$

After setting $\tilde{\theta} = \theta + \phi$, we can rewrite this equivalence relation in the form

$$(\phi, \tilde{\theta}, r, \tilde{u}, \tilde{v}) \sim (\phi + \pi, \tilde{\theta}, r, -\tilde{u} - \tilde{v}).$$

The variable $\tilde{\theta}$ now no longer plays a role in the gluing. After replacing $\mathbb{R}^+$ by $\mathbb{R}$ and $\dot{D}^2$ by $\mathbb{R}^2$, we see $\mathcal{S}^4_+$ is diffeomorphic to $S^1 \times S^1 \times \mathbb{R}^3$ modulo the gluing relation

$$(\phi, \tilde{\theta}, x^1, x^2, x^3) \sim (\phi + \pi, \tilde{\theta}, x^1, -x^2, -x^3). \qquad (13.4.\mathrm{d})$$

The gluing of Equation (13.4.d) defines the total space of the bundle $\mathbb{1} \oplus \mathbb{L} \oplus \mathbb{L}$ over (S^1, ϕ). Since $\mathbb{L} \oplus \mathbb{L}$ is just the trivial 2-plane bundle $\mathbb{1} \oplus \mathbb{1}$, we obtain finally that $\mathcal{S}^4_{+,0}$ is diffeomorphic to $S^1 \times S^1 \times \mathbb{R}^3$. □

13.4.3 THE INVARIANTS ψ AND Ψ WHEN RANK$(\rho) = 2$. We suppose the Ricci tensor is non-degenerate for the remainder of this section and let ρ^{ij} be the inverse of the matrix ρ_{ij}. Linear equivalence and affine equivalence are the same in this setting by Theorem 13.5. Since contraction of an upper index against a lower index is invariant under the action of $GL(2, \mathbb{R})$, the following are $GL(2, \mathbb{R})$ invariants:

$$\rho_{v,ij} := \Gamma_{ik}{}^{\ell}\Gamma_{j\ell}{}^{k}, \quad \psi := \text{Tr}_{\rho}\{\rho_v\} = \rho^{ij}\rho_{v,ij}, \quad \Psi := \det(\rho_v)/\det(\rho).$$

We refer to Brozos-Vázquez, García-Río, and Gilkey [6] for the proof of the following result as it is beyond the scope of this book.

Theorem 13.36 *Let $\mathcal{M}$ and $\tilde{\mathcal{M}}$ be two Type $\mathcal{A}$ connections such that ρ_{∇} and $\rho_{\tilde{\nabla}}$ are non-degenerate and have the same signature. Then $\mathcal{M}$ and $\tilde{\mathcal{M}}$ are affine equivalent if and only if we have that $(\psi, \Psi)(\mathcal{M}) = (\psi, \Psi)(\tilde{\mathcal{M}})$.*

We show the image of (ψ, Ψ) below in Figure 13.4; the region on the far right is the moduli space for positive definite Ricci tensor, the central region is the moduli space for indefinite Ricci tensor, and the region on the left the moduli space for negative definite Ricci tensor. The left boundary curve between negative definite and indefinite Ricci tensors is σ_{ℓ} (given in red) and the right boundary curve between indefinite and positive definite Ricci tensors is $\sigma_r(t)$ (given in blue) where

$$\sigma_{\ell}(t) := (-4t^2 - t^{-2} + 2, 4t^4 - 4t^2 + 2),$$
$$\sigma_r(t) := (4t^2 + t^{-2} + 2, 4t^4 + 4t^2 + 2).$$

Figure 13.4: Moduli spaces of Type $\mathcal{A}$ models with $\det(\rho) \neq 0$.

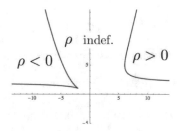

Note that although (ψ, Ψ) is 1-1 on each of the three cases separately, the images intersect along the smooth curves σ_{ℓ} and σ_r.

13.4.4 LINEAR EQUIVALENCE FOR THE MANIFOLDS $\mathcal{M}_1^2(\cdot)$. If we have that $\mathcal{M} = \mathcal{M}_1^2(a_1, a_2)$, then $\mathcal{Q}(\mathcal{M}) = \text{span}\{e^{L_1}, e^{L_2}, e^{L_3}\}$. Suppose that $\{L_i, L_j\}$ are linearly independent for $i \neq j$. Let σ be a permutation of the integers $\{1, 2, 3\}$. Introduce new coordinates

$y_\sigma^1 = L_{\sigma(1)}(x^1, x^2)$ and $y_\sigma^2 := L_{\sigma(2)}(x^1, x^2)$. Expand $L_{\sigma(3)}(x^1, x^2) = a_{1,\sigma} y_\sigma^1 + a_{2,\sigma} y_\sigma^2$ to express

$$\mathcal{Q}(\mathcal{M}) = \text{span}\{e^{y_\sigma^1}, e^{y_\sigma^2}, e^{a_{1,\sigma} y_\sigma^1 + a_{2,\sigma} y_\sigma^2}\}.$$

This structure is defined by the pair $(a_{1,\sigma}, a_{2,\sigma})$; there are, generically, six such pairs that give rise to the same affine structure up to linear equivalence. If $\mathcal{M}_1^2(a_1, a_2)$ is linearly equivalent to $\mathcal{M}_1^2(\tilde{a}_1, \tilde{a}_2)$, then we shall write $(a_1, a_2) \sim (\tilde{a}_1, \tilde{a}_2)$. This means that there exists $T \in \mathbb{R}^2$ so $T \, \text{span}\{e^{x^1}, e^{x^2}, e^{a_1 x^1 + a_2 x^2}\} = \text{span}\{e^{\tilde{x}^1}, e^{\tilde{x}^2}, e^{\tilde{a}_1 \tilde{x}^1 + \tilde{a}_2 \tilde{x}^2}\}$. Suppose that $L_1 = x^1$, $L_2 = x^2$, and $L_3 = a_1 x^1 + a_2 x^2$. Let σ_{ijk} be the permutation $1 \to i, 2 \to j, 3 \to k$. We have

$$\begin{aligned}
\sigma_{123}: & \quad y^1 = L_1, \quad y^2 = L_2, \quad L_3 = a_1 y^1 + a_2 y^2, & (a_1, a_2) \sim (a_1, a_2). \\
\sigma_{213}: & \quad y^1 = L_2, \quad y^2 = L_1, \quad L_3 = a_2 y^1 + a_1 y^2, & (a_1, a_2) \sim (a_2, a_1). \\
\sigma_{132}: & \quad y^1 = L_1, \quad y^2 = L_3, \quad L_2 = -\tfrac{a_1}{a_2} y^1 + \tfrac{1}{a_2} y^2, & (a_1, a_2) \sim (-\tfrac{a_1}{a_2}, \tfrac{1}{a_2}). \\
\sigma_{321}: & \quad y^1 = L_3, \quad y^2 = L_2, \quad L_1 = \tfrac{1}{a_1} y^1 - \tfrac{a_2}{a_1} y^2, & (a_1, a_2) \sim (\tfrac{1}{a_1}, -\tfrac{a_2}{a_1}). \\
\sigma_{231}: & \quad y^1 = L_2, \quad y^2 = L_3, \quad L_1 = -\tfrac{a_2}{a_1} y^1 + \tfrac{1}{a_1} y^2, & (a_1, a_2) \sim (-\tfrac{a_2}{a_1}, \tfrac{1}{a_1}). \\
\sigma_{312}: & \quad y^1 = L_3, \quad y^2 = L_1, \quad L_2 = \tfrac{1}{a_2} y^1 - \tfrac{a_1}{a_2} y^2, & (a_1, a_2) \sim (\tfrac{1}{a_2}, -\tfrac{a_1}{a_2}).
\end{aligned}$$

We observe that since ψ and Ψ are linear invariants, they are constant under the action of the group of permutations s_3. Although generically s_3 acts without fixed points, there are degenerate cases where the action is not fixed point free.

If $\det(\rho) > 0$ and $\text{Tr}\{\rho\} < 0$, then ρ is negative definite; if $\det(\rho) > 0$ and $\text{Tr}\{\rho\} > 0$, then ρ is positive definite; if $\det(\rho) < 0$, then ρ is indefinite. The six lines

$$\{x = 0, \; x = -1, \; y = 0, \; y = -1, \; x + y = 1, \; x = y\}$$

are given in black below; they further divide the regions where ρ is negative definite (light blue), ρ is positive definite (yellow), and ρ is indefinite (green); the three regions in different colors can be further divided into six regions under the action of s_3 under the picture given below.

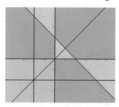

13.4.5 LINEAR EQUIVALENCE FOR THE MANIFOLDS $\mathcal{M}_2^2(\cdot)$.

We have that $\mathcal{Q}(\mathcal{M}) = \text{span}\{e^{L_1} \cos(L_2), e^{L_1} \sin(L_2), e^{L_3}\}$. We set $\mathcal{M} = \mathcal{M}_2^2(b_1, b_2)$ where $b_1 \neq 1$ and $(b_1, b_2) \neq (0, 0)$. We have $b_1 > 1$ corresponds to ρ positive definite and $b_1 < 1$ corresponds to ρ indefinite; (b_1, b_2) and $(\tilde{b}_1, \tilde{b}_2)$ are linearly equivalent if and only if $b_1 = \tilde{b}_1$ and $b_2 = \pm \tilde{b}_2$.

CHAPTER 14

The Geometry of Type $\mathcal{B}$ Models

By Theorem 12.49, a locally affine homogeneous surface falls into one of three types. The Type $\mathcal{A}$ models are precisely those where there exists an effective Abelian 2-dimensional Lie subalgebra of the Lie algebra of affine Killing vector fields. They are modeled on the geometry of a left-invariant affine connection on $\mathbb{R}^2$ and the geometry of these surfaces is discussed at some length in Chapter 13. The Type $\mathcal{B}$ models are those where there exists an effective non-Abelian 2-dimensional Lie subalgebra of the Lie algebra of affine Killing vector fields. These are modeled on the geometry of a left-invariant affine connection on the $ax + b$ group. The Christoffel symbols take the form

$$\Gamma_{ij}{}^k(x^1, x^2) = \frac{1}{x^1} A_{ij}{}^k \quad \text{for} \quad A_{ij}{}^k \in \mathbb{R}.$$

Chapter 14 is devoted to the study of their geometry.

14.1 TYPE $\mathcal{B}$: DISTINGUISHED GEOMETRIES

Here is a brief outline to this section. We begin in Definition 14.1 by defining the fundamental structures of interest. In Lemma 14.2 of Section 14.1.1, we discuss how the Christoffel symbols transform under a shear

$$(x^1, x^2) \rightarrow (x^1, ax^2 + bx^1).$$

In Lemma 14.3 of Section 14.1.2, we determine the Ricci tensor of the non-flat models of Definition 14.1; the spaces $\mathcal{Q}(\cdot)$ are treated in Lemma 14.4 of Section 14.1.3 and in Lemma 14.5 of Section 14.1.4. Eigenspaces of the affine quasi-Einstein equation for eigenvalues $\mu \notin \{0, -1\}$ are studied in Lemma 14.6 of Section 14.1.5. The topology of the space of flat non-trivial Type $\mathcal{B}$ models is presented in Theorem 14.7 in Section 14.1.6; by contrast with the Type $\mathcal{A}$ setting, it is not smooth but consists of three smooth closed surfaces which intersect transversally. In Theorem 14.8 of Section 14.1.7, we perform similar analysis for the Type $\mathcal{B}$ models with $\rho_s = 0$ but $\rho_a \neq 0$. In Theorem 14.9 of Section 14.1.8, we use the parameterizations of Theorem 14.7 and Theorem 14.8 to show that any flat geometry is linearly equivalent to one of the models $\mathcal{N}_i^6(\cdot)$ and that any model which is not flat but with $\rho_s = 0$ is linearly equivalent to one of the models $\mathcal{N}_3^2(0, c, \pm)$ for $c \neq 0$ or $\mathcal{N}_2^2(c)$. We also show that $\mathcal{N}$ is strongly projectively flat but not flat if it is linearly equivalent to one of the models $\mathcal{N}_i^4(\cdot), \mathcal{N}_3^3, \mathcal{N}_4^3,$ or $\mathcal{N}_1^2(\cdot)$. In Lemma 14.11 of Section 14.1.9, we establish some useful affine immersion and embedding results. In Lemma 14.12

of Section 14.1.10, we treat Yamabe solitons, and in Lemma 14.13 of Section 14.1.11, we examine affine gradient Ricci solitons.

In analogy to the Type $\mathcal{A}$ models which were defined in Equation (13.2.a), we let $\mathcal{N}(\xi_1, \xi_2, \xi_3, \xi_4, \xi_5, \xi_6)$ be the Type $\mathcal{B}$ model defined by the Christoffel symbols

$$\left\{ \Gamma_{11}{}^1 = \frac{\xi_1}{x^1}, \ \Gamma_{11}{}^2 = \frac{\xi_2}{x^1}, \ \Gamma_{12}{}^1 = \frac{\xi_3}{x^1}, \ \Gamma_{12}{}^2 = \frac{\xi_4}{x^1}, \ \Gamma_{22}{}^1 = \frac{\xi_5}{x^1}, \ \Gamma_{22}{}^2 = \frac{\xi_6}{x^1} \right\} .$$

Definition 14.1   We define the following Type $\mathcal{B}$ structures.

$\mathcal{N}_i^6$: $\mathcal{N}_0^6 := \mathcal{N}(0, 0, 0, 0, 0, 0)$, $\mathcal{N}_1^6(\pm) := \mathcal{N}(1, 0, 0, 0, \pm 1, 0)$,

$\qquad\ \mathcal{N}_2^6(c_1) := \mathcal{N}(c_1 - 1, 0, 0, c_1, 0, 0)$, $\mathcal{N}_3^6 := \mathcal{N}(-2, 1, 0, -1, 0, 0)$,

$\qquad\ \mathcal{N}_4^6 := \mathcal{N}(0, 1, 0, 0, 0, 0)$, $\mathcal{N}_5^6 := \mathcal{N}(-1, 0, 0, 0, 0, 0)$,

$\qquad\ \mathcal{N}_6^6(c_2) := \mathcal{N}(c_2, 0, 0, 0, 0, 0)$,

$\qquad\ $ where $c_1 \neq 0$ and $c_2 \notin \{0, -1\}$.

$\mathcal{N}_i^4$: $\mathcal{N}_1^4(c_2) := \mathcal{N}(2c_2, 1, 0, c_2, 0, 0)$,

$\qquad\ \mathcal{N}_2^4(\kappa, \theta) := \mathcal{N}(2\kappa + \theta - 1, 0, 0, \kappa, 0, 0)$, $\mathcal{N}_3^4(c_1) := \mathcal{N}(2c_1 - 1, 0, 0, c_1, 0, 0)$,

$\qquad\ $ where $c_1 \neq 0$, $c_2 \notin \{0, -1\}$, $\theta \neq 0$ and $\kappa(\kappa + \theta) \neq 0$.

$\mathcal{N}_i^3$: $\mathcal{N}_1^3(\pm) := \mathcal{N}(-\frac{3}{2}, 0, 0, -\frac{1}{2}, \mp\frac{1}{2}, 0)$, $\mathcal{N}_2^3(c) := \mathcal{N}(-\frac{3}{2}, 0, 1, -\frac{1}{2}, c, 2)$,

$\qquad\ \mathcal{N}_3^3 := \mathcal{N}(-1, 0, 0, -1, -1, 0)$, $\mathcal{N}_4^3 := \mathcal{N}(-1, 0, 0, -1, 1, 0)$,

$\qquad\ $ where $c \in \mathbb{R}$.

$\mathcal{N}_i^2$: $\mathcal{N}_1^2(v, \pm) := \mathcal{N}(1 + 2v, 0, 0, v, \pm 1, 0)$, $v \notin \{0, -1\}$,

$\qquad\ \mathcal{N}_2^2(c) := \mathcal{N}(0, c, 1, 0, 0, 1)$, $c \in \mathbb{R}$,

$\qquad\ \mathcal{N}_3^2(a, c, \pm) := \mathcal{N}(\frac{1}{2}(a^2 + 4a \mp 2c^2 + 2), c, 0, \frac{1}{2}(a^2 + 2a \mp 2c^2), \pm 1, \pm 2c)$,

$\qquad\ $ where $(a, c) \neq (0, 0)$ and $c \geq 0$.

$\mathcal{N}_3^3$ is the Lorentzian hyperbolic plane and $\mathcal{N}_4^3$ is the hyperbolic plane; the connections are the Levi-Civita connections of the metrics $ds^2 = (x^1)^{-2}\{(dx^1)^2 \mp (dx^2)^2\}$. The notation is chosen so that (see Theorem 14.16) $\dim(\mathfrak{K}(\mathcal{N}_i^v)) = v$ except for $\mathcal{N}_3^2(0, \frac{3}{\sqrt{2}}, +)$, which is equivalent to $\mathcal{N}_2^3(\frac{1}{2})$.

14.1.1 LINEAR EQUIVALENCE.

The natural gauge group for the Type $\mathcal{A}$ geometries was the general linear group. The corresponding gauge group in the Type $\mathcal{B}$ setting is much smaller. If $T(x^1, x^2) = (ax^1, ax^2 + b)$ for $a \neq 0$ and if $\mathcal{N}$ is a Type $\mathcal{B}$ geometry, then $T^*\mathcal{N} = \mathcal{N}$. Thus these are homogeneous geometries. The $ax + b$ group has another action on $\mathbb{R}^+ \times \mathbb{R}$ by means of the shear transformation $T(x^1, x^2) = (x^1, ax^2 + bx^1)$ and is the appropriate gauge group in this setting. We say that two Type $\mathcal{B}$ geometries are *linearly equivalent* if there exists a shear intertwining them. This is a much smaller gauge group than $\mathrm{GL}(2,\mathbb{R})$ and for that reason the Type $\mathcal{B}$ geometries are much richer. The following is a useful result which follows by a direct computation.

Lemma 14.2   *Let $(y^1, y^2) = (x^1, a^{-1}(x^2 - bx^1))$ be a change of variables which defines a shear. Then*

$$dy^1 = dx^1, \quad dy^2 = a^{-1}(dx^2 - b dx^1), \quad \partial_{y^1} = \partial_{x^1} + b\partial_{x^2}, \quad \partial_{y^2} = a\partial_{x^2},$$
$$\,^y A_{11}{}^1 = \,^x A_{11}{}^1 + 2b\,^x A_{12}{}^1 + b^2\,^x A_{22}{}^1,$$
$$\,^y A_{11}{}^2 = \tfrac{1}{a}\{\,^x A_{11}{}^2 + b(2\,^x A_{12}{}^2 - \,^x A_{11}{}^1) + b^2(\,^x A_{22}{}^2 - 2\,^x A_{12}{}^1) - b^3\,^x A_{22}{}^1\},$$
$$\,^y A_{12}{}^1 = a(\,^x A_{12}{}^1 + b\,^x A_{22}{}^1),$$
$$\,^y A_{12}{}^2 = \,^x A_{12}{}^2 + b\,^x A_{22}{}^2 - b(\,^x A_{12}{}^1 + b\,^x A_{22}{}^1),$$
$$\,^y A_{22}{}^1 = a^2\,^x A_{22}{}^1,$$
$$\,^y A_{22}{}^2 = a(\,^x A_{22}{}^2 - b\,^x A_{22}{}^1).$$

14.1.2 THE RICCI TENSOR.   To clear denominators, we will let $\tilde{\rho} := (x^1)^2 \rho$ throughout this section. A direct computation shows

$$\tilde{\rho}_{11} = (1 + A_{11}{}^1 - A_{12}{}^2)A_{12}{}^2 + A_{11}{}^2(-A_{12}{}^1 + A_{22}{}^2),$$
$$\tilde{\rho}_{12} = A_{12}{}^1 A_{12}{}^2 - A_{11}{}^2 A_{22}{}^1 + A_{22}{}^2,$$
$$\tilde{\rho}_{21} = A_{12}{}^1(-1 + A_{12}{}^2) - A_{11}{}^2 A_{22}{}^1,$$
$$\tilde{\rho}_{22} = -(A_{12}{}^1)^2 + (-1 + A_{11}{}^1 - A_{12}{}^2)A_{22}{}^1 + A_{12}{}^1 A_{22}{}^2.$$

$$(14.1.a)$$

In particular, note that the Ricci tensor is not in general symmetric. We make a direct computation to establish the following result.

Lemma 14.3   *The Ricci tensor of the geometries $\mathcal{N}_i^6(\cdot)$ vanishes. These geometries are flat. Moreover:*

$\mathcal{N}_i^4:\quad \tilde{\rho}_{\mathcal{N}_1^4(c_2)} = c_2(1 + c_2)dx^1 \otimes dx^1, \qquad\qquad \tilde{\rho}_{\mathcal{N}_2^4(\kappa,\theta)} = \kappa(\kappa + \theta)dx^1 \otimes dx^1,$

$\qquad\quad \tilde{\rho}_{\mathcal{N}_3^4(c_1)} = c_1^2 dx^1 \otimes dx^1.$

$\mathcal{N}_i^3:\quad \tilde{\rho}_{\mathcal{N}_1^3(\pm)} = \pm dx^2 \otimes dx^2, \qquad\qquad\qquad \tilde{\rho}_{\mathcal{N}_2^3(c)} = \begin{pmatrix} 0 & \frac{3}{2} \\ -\frac{3}{2} & 1 - 2c \end{pmatrix},$

$\qquad\quad \tilde{\rho}_{\mathcal{N}_3^3} = -(dx^1)^2 + (dx^2)^2, \qquad\qquad \tilde{\rho}_{\mathcal{N}_4^3} = -(dx^1)^2 - (dx^2)^2.$

$\mathcal{N}_i^2:\quad \tilde{\rho}_{\mathcal{N}_1^2(v,\pm)} = v(2 + v)(dx^1)^2 \pm v(dx^2)^2, \qquad \tilde{\rho}_{\mathcal{N}_2^2(c)} = \begin{pmatrix} 0 & 1 \\ -1 & 0 \end{pmatrix},$

$\qquad\quad \tilde{\rho}_{\mathcal{N}_3^2(a,c,\pm)} = \begin{pmatrix} \frac{1}{2}a\left(a^2 + 4a \mp 2c^2 + 4\right) & \pm c \\ \mp c & \pm a \end{pmatrix}.$

14.1.3 THE AFFINE QUASI-EINSTEIN EQUATION. All the geometries of Definition 14.1 with the exception of $\mathcal{N}_1^3(\pm)$, $\mathcal{N}_2^3(c)$, $\mathcal{N}_2^2(c)$, and $\mathcal{N}_3^2(a, c, \pm)$ are strongly projectively flat; we shall show in Theorem 14.9 that, up to linear equivalence, there are no other strongly projectively flat geometries.

Lemma 14.4 *We have:*

$\mathcal{Q}(\mathcal{N}_i^6)$: $\mathcal{Q}(\mathcal{N}_0^6) = \text{span}\{\text{⫘}, x^1, x^2\}$, $\qquad \mathcal{Q}(\mathcal{N}_1^6(\pm)) = \text{span}\{\text{⫘}, x^2, (x^1)^2 \pm (x^2)^2\}$,

$\qquad \mathcal{Q}(\mathcal{N}_2^6(c_1)) = \text{span}\{\text{⫘}, (x^1)^{c_1}, (x^1)^{c_1} x^2\}$, $\quad \mathcal{Q}(\mathcal{N}_3^6) = \text{span}\{\text{⫘}, \frac{1}{x^1}, \frac{x^2}{x^1} + \log(x^1)\}$,

$\qquad \mathcal{Q}(\mathcal{N}_4^6) = \text{span}\{\text{⫘}, x^1, x^2 + x^1 \log(x^1)\}$, $\quad \mathcal{Q}(\mathcal{N}_5^6) = \text{span}\{\text{⫘}, \log(x^1), x^2\}$,

$\qquad \mathcal{Q}(\mathcal{N}_6^6(c_2)) = \text{span}\{\text{⫘}, (x^1)^{1+c_2}, x^2\}$.

$\mathcal{Q}(\mathcal{N}_i^4)$: $\mathcal{Q}(\mathcal{N}_1^4(c_2)) = (x^1)^{c_2} \mathcal{Q}(\mathcal{N}_4^6)$,

$\qquad \mathcal{Q}(\mathcal{N}_2^4(\kappa, \theta)) = (x^1)^\kappa \text{span}\{\text{⫘}, x^2, (x^1)^\theta\}$, $\quad \mathcal{Q}(\mathcal{N}_3^4(c_1)) = (x^1)^{c_1} \mathcal{Q}(\mathcal{N}_5^6)$.

$\mathcal{Q}(\mathcal{N}_i^3)$: $\mathcal{Q}(\mathcal{N}_1^3(\pm)) = \mathcal{Q}(\mathcal{N}_2^3(c)) = \{0\}$, *for* $c \neq \frac{1}{2}$, $\mathcal{Q}(\mathcal{N}_2^3(\frac{1}{2})) = \mathbb{R} \cdot \text{⫘}$,

$\qquad \mathcal{Q}(\mathcal{N}_3^3) = \text{span}\{\frac{1}{x^1}, \frac{x^2}{x^1}, \frac{(x^2)^2 - (x^1)^2}{x^1}\}$, $\qquad \mathcal{Q}(\mathcal{N}_4^3) = \text{span}\{\frac{1}{x^1}, \frac{x^2}{x^1}, \frac{(x^2)^2 + (x^1)^2}{x^1}\}$.

$\mathcal{Q}(\mathcal{N}_i^2)$: $\mathcal{Q}(\mathcal{N}_1^2(v, \pm)) = (x^1)^v \mathcal{Q}(\mathcal{N}_1^6(\pm))$, $\qquad \mathcal{Q}(\mathcal{N}_2^2(c)) = \mathbb{R} \cdot \text{⫘}$.

$\qquad$ *If* $a \neq 0$, *then* $\mathcal{Q}(\mathcal{N}_3^2(a \neq 0, c, \pm)) = \{0\}$.

$\qquad$ *If* $a = 0$, *then* $\mathcal{Q}(\mathcal{N}_3^2(0, c, \pm)) = \mathbb{R} \cdot \text{⫘}$.

Proof. By Lemma 12.14, $\dim(\mathcal{Q}(\cdot)) \leq 3$. With the exception of the geometries $\mathcal{N}_1^3(\pm)$, $\mathcal{N}_2^3(c)$, $\mathcal{N}_2^2(c)$, and $\mathcal{N}_3^2(a, c, \pm)$, we need only verify that the functions given belong to $\mathcal{Q}(\cdot)$; they span for dimensional reasons. Theorem 12.34 then implies all these geometries are strongly projectively flat. We apply Lemma 12.53 to see that if $\mathcal{Q}(\mathcal{N})$ is non-trivial, then $(x^1)^\alpha \in \mathcal{Q}_\mathbb{C}(\mathcal{N})$ for some $\alpha \in \mathbb{C}$. This fails for the geometries $\mathcal{N}_1^3(\pm)$ and $\mathcal{N}_2^3(c)$ with $c \neq \frac{1}{2}$. This shows that $\mathcal{Q}(\cdot) = \{0\}$ in these instances. In the geometry $\mathcal{N}_2^2(c)$ as well as $\mathcal{N}_2^3(\frac{1}{2})$, the Ricci tensor is alternating; we verify ⫘ spans $\mathcal{Q}(\cdot)$. In the geometry $\mathcal{N}_3^2(a, c, \pm)$, a similar computation pertains; the Ricci tensor is alternating if and only if $a = 0$. $\qquad\square$

14.1.4 TYPE $\mathcal{B}$ GEOMETRIES WITH DIM($\mathcal{Q}$) = 1. We use Lemma 12.32 to see that if $\dim(\mathcal{Q}(\mathcal{N})) = 3$, then $\mathcal{N}$ is strongly projectively flat; we will show in Theorem 14.9 that if $\mathcal{N}$ is strongly projectively flat, then it appears among the manifolds of Definition 14.1. We refer to Brozos-Vázquez et al. [9] for a slightly different discussion.

Lemma 14.5 *Suppose $\mathcal{N}$ is a Type $\mathcal{B}$ geometry with $\dim(\mathcal{Q}(\mathcal{N})) = 1$ and $\rho_s \neq 0$. Then one of the following possibilities pertains.*

1. *$\mathcal{N}$ is linearly equivalent to $\mathcal{N}(1 + 2d + c^2 \varepsilon, c, 0, d, \varepsilon, 2c\varepsilon)$ for some (c, d) where we have that $c \neq 0$, $c^2 \varepsilon + d \neq 0$, and $\varepsilon = \pm 1$. We have $\mathcal{Q}(\mathcal{N}) = \text{span}\{(x^1)^{d + c^2 \varepsilon}\}$ and*

$$\tilde{\rho} = \begin{pmatrix} (d + 2)(\varepsilon c^2 + d) & c\varepsilon \\ -c\varepsilon & \varepsilon(\varepsilon c^2 + d) \end{pmatrix}.$$

2. $\mathcal{N}$ is linearly equivalent to $\mathcal{N}(0, b, 1, a, 0, 1)$ for some (a, b) with $a \neq 0$. We then have that $\mathcal{Q}(\mathcal{N}) = \mathrm{span}\{(x^1)^a\}$ and

$$\tilde{\rho} = \begin{pmatrix} a - a^2 & a + 1 \\ a - 1 & 0 \end{pmatrix}.$$

Proof. Let $\Gamma_{ij}{}^k = (x^1)^{-1} A_{ij}{}^k$ be a Type $\mathcal{B}$ geometry. We set $\mu = -1$. We use Lemma 12.53 and distinguish cases.

Case 1. Suppose $A_{22}{}^1 \neq 0$. We can rescale x^2 to assume $A_{22}{}^1 = \varepsilon = \pm 1$. We use Lemma 14.2 to apply a shear to assume $A_{12}{}^1 = 0$. This fixes the gauge. We renormalize the affine quasi-Einstein equation by multiplying by $(x^1)^{2-\alpha}$ to simplify expressions. We set $f = (x^1)^\alpha$ to obtain

$$\tilde{\mathfrak{Q}}_{11} : 0 = (1 + A_{11}{}^1) A_{12}{}^2 - (A_{12}{}^2)^2 + A_{22}{}^2 A_{11}{}^2 + \alpha(-1 - A_{11}{}^1 + \alpha),$$
$$\tilde{\mathfrak{Q}}_{12} : 0 = \tfrac{1}{2}(A_{22}{}^2 - 2A_{11}{}^2\varepsilon), \qquad \tilde{\mathfrak{Q}}_{22} : 0 = (-1 + A_{11}{}^1 - A_{12}{}^2 - \alpha)\varepsilon.$$

We set $A_{11}{}^2 = c$ and $A_{12}{}^2 = d$ and solve these relations to obtain the defining relations of Assertion 1:

$$
\begin{aligned}
& A_{11}{}^1 = 1 + 2d + c^2\varepsilon, && A_{11}{}^2 = c, && A_{12}{}^1 = 0, \\
& A_{12}{}^2 = d, && A_{22}{}^1 = \varepsilon, && A_{22}{}^2 = 2c\varepsilon, \\
& \tilde{\rho} = \begin{pmatrix} (2 + d)(d + c^2\varepsilon) & c\varepsilon \\ -c\varepsilon & \varepsilon(d + c^2\varepsilon) \end{pmatrix}, && \alpha = d + c^2\varepsilon.
\end{aligned}
\tag{14.1.b}
$$

To ensure $\rho_s \neq 0$, we assume $c^2\varepsilon + d \neq 0$. We must show there are no other solutions if Equation (14.1.b) holds. Since α is unique, any other solution must have the form

$$f = (x^1)^\alpha \log(x^1) \quad \text{or} \quad f = (x^1)^\alpha (a_1 x^1 + x^2).$$

1. If $f = (x^1)^\alpha \log(x^1)$, then $\tilde{\mathfrak{Q}}_{22} : 0 = -\varepsilon$ which is impossible as $\varepsilon = \pm 1$.
2. If $f = (x^1)^\alpha (a_1 x^1 + x^2)$, then $\tilde{\mathfrak{Q}}_{12} : 0 = c^2\varepsilon x^1$ so $c = 0$. Conversely, if $c = 0$, then $(x^1)^\alpha x^2 \in \mathcal{Q}(\mathcal{N})$. Consequently, the restriction that $c \neq 0$ is essential. This establishes the possibility discussed in Assertion 1.

Case 2. We assume $A_{22}{}^1 = 0$ but $A_{12}{}^1 \neq 0$. We can rescale x^2 to assume $A_{12}{}^1 = 1$ and then normalize the gauge by performing a shear to assume $A_{11}{}^1 = 0$. Set $f = (x^1)^\alpha$ and $\mu = -1$; we obtain $\tilde{\mathfrak{Q}}_{22}: 0 = -1 + A_{22}{}^2$. We set $A_{22}{}^2 = 1$ and obtain that $\tilde{\mathfrak{Q}}_{12}: 0 = A_{12}{}^2 - \alpha$. This implies that $\alpha = A_{12}{}^2$. This yields $\tilde{\mathfrak{Q}}_{11} = 0$. We set $A_{11}{}^2 = b$ and $A_{12}{}^2 = a$ to obtain the defining relations of Assertion 2:

$$
\begin{aligned}
& A_{11}{}^1 = 0, && A_{11}{}^2 = b, && A_{12}{}^1 = 1, \\
& A_{12}{}^2 = a, && A_{22}{}^1 = 0, && A_{22}{}^2 = 1, \\
& \tilde{\rho} = \begin{pmatrix} a - a^2 & a + 1 \\ a - 1 & 0 \end{pmatrix}, && \alpha = a.
\end{aligned}
\tag{14.1.c}
$$

To ensure $\rho_s \neq 0$, we have $a \neq 0$. We impose the relations of Equation (14.1.c). We suppose there exists another solution f and argue for a contradiction. The exponent is unique; therefore the other possibilities are:

$$f = (x^1)^\alpha \log(x^1) \quad \text{or} \quad f = (x^1)^\alpha (a_1 x^1 + x^2).$$

1. If $f = (x^1)^\alpha \log(x^1)$, then $\tilde{\mathfrak{Q}}_{12} : 0 = -1$ so this is impossible.
2. If $f = (x^1)^\alpha (a_1 x^1 + x^2)$, then $\tilde{\mathfrak{Q}}_{22} : 0 = -x^1$ so this is impossible.

Case 3. We assume $A_{22}{}^1 = 0$ and $A_{12}{}^1 = 0$. We obtain $\tilde{\mathfrak{Q}}_{12} : 0 = \frac{1}{2} A_{22}{}^2$ so $A_{22}{}^2 = 0$ as well. We will show subsequently in Theorem 14.17 that this implies that $\mathcal{N}$ is linearly equivalent to $\mathcal{N}_i^4(\cdot)$ so $\dim(\mathcal{Q}(\mathcal{N})) = 3$ and this case does not occur. Nevertheless, it is worth giving a direct argument. We suppose $A_{12}{}^1 = A_{22}{}^1 = A_{22}{}^2 = 0$. The remaining equation becomes

$$\tilde{\mathfrak{Q}}_{11} : \ 0 = -(A_{12}{}^2 - \alpha)(-1 - A_{11} + A_{12}{}^2 + \alpha).$$

1. If $A_{11}{}^1 \neq 2A_{12}{}^2 - 1$, then there are two possible values for α. Thus, $\dim(\mathcal{Q}(\mathcal{N})) > 1$, and this case is impossible.
2. If $A_{11}{}^1 = 2A_{12}{}^2 - 1$, then $(x^1)^\alpha (a_0 + a_1 \log(x^1))$ is a solution for $\alpha = A_{12}{}^2$ and a_0 and a_1 arbitrary. Thus again $\dim(\mathcal{Q}(\mathcal{N})) > 1$ so this case is impossible as well. $\qquad \square$

14.1.5 OTHER EIGENVALUES OF THE AFFINE QUASI-EINSTEIN EQUATION.

We shall assume $\mathcal{N}$ is not flat. If $\rho_s = 0$, then $\rho_a \neq 0$ and it follows from Theorem 14.9 that $\mathcal{N} = \mathcal{N}_2^2(\cdot)$ or $\mathcal{N} = \mathcal{N}_3^2(\cdot)$. We suppose $\mu \neq 0$ since we will study Yamabe solitons in Lemma 14.12. We also suppose $\mu \neq -1$ since we examined that setting in the previous section. By Lemma 12.32, $\dim(E(\mu, \mathcal{N})) = 3$ if and only if $\rho_s = 0$. Consequently, $\dim(E(\mu, \mathcal{N})) \leq 2$. By Lemma 12.31, $\mathbb{1} \notin E(\mu, \mathcal{N})$. Thus we will examine $(x^1)^\alpha$ for $\alpha \neq 0$. We shall follow the same argument used to prove Lemma 14.5. We refer to Brozos-Vázquez et al. [9] for a slightly different discussion.

Lemma 14.6   *Suppose $\mathcal{N}$ is a Type $\mathcal{B}$ model with $\rho_s \neq 0$, with $\dim(E(\mu, \mathcal{N})) \geq 1$, and with $\mu \neq 0, -1$. Then one of the following possibilities pertains.*

1. *$\mathcal{N}$ is linearly equivalent to $\mathcal{N} = \mathcal{N}(a, c, 0, b, \varepsilon, 2c\varepsilon)$ for $a - b - 1 \neq 0$,*

$$\mu = \frac{-a^2 + 2ab - b^2 + 2b + 2c^2\varepsilon + 1}{(a - b - 1)^2}, \ \alpha = \frac{a^2 - 2ab + b^2 - 2b - 2c^2\varepsilon - 1}{a - b - 1}, \ \text{and } E(\mu, \mathcal{N}) = \text{span}\{(x^1)^\alpha\}.$$

2. *$\mathcal{N}$ is linearly equivalent to $\mathcal{N}_i^4(\cdot)$. We have $A_{12}{}^1 = A_{22}{}^1 = A_{22}{}^2 = 0$ and $\tilde{\rho} = (1 + A_{11}{}^1 - A_{12}{}^2)A_{12}{}^2 dx^1 \otimes dx^1 \neq 0$. Denote the discriminant by $D := (A_{11}{}^1)^2 + 4A_{11}{}^1 A_{12}{}^2 \mu + 2A_{11}{}^1 - 4(A_{12}{}^2)^2 \mu + 4A_{12}{}^2 \mu + 1$.*
   (a) *If $D \neq 0$, then $E_{\mathbb{C}}(\mu, \mathcal{N}) = \text{span}\{(x^1)^{\frac{1}{2}(1 + A_{11}{}^1 + \sqrt{D})}, (x^1)^{\frac{1}{2}(1 + A_{11}{}^1 - \sqrt{D})}\}$. If $D < 0$, we must take real and imaginary parts to get real solutions.*

(b) If $D = 0$, then $\mu = \frac{(A_{11}{}^1)^2 + 2A_{11}{}^1 + 1}{4A_{12}{}^2(-A_{11}{}^1 + A_{12}{}^2 - 1)}$ and
$E(\mu, \mathcal{N}) = \mathrm{span}\{(x^1)^{\frac{1}{2}(1+A_{11}{}^1)}, (x^1)^{\frac{1}{2}(1+A_{11}{}^1)} \log(x^1)\}$.

Proof. We distinguish cases. We assume $\mu \neq 0, -1$ and $\alpha \neq 0$.

Case 1. Suppose $A_{22}{}^1 \neq 0$. We rescale x^2 to assume $A_{22}{}^1 = \varepsilon = \pm 1$. Use Lemma 14.2 to choose a shear to assume that $A_{12}{}^1 = 0$; this fixes the gauge. We obtain

$$\ddot{\mathfrak{Q}}_{\mu,12} : 0 = A_{11}{}^2 \mu \varepsilon - \frac{A_{22}{}^2 \mu}{2}, \quad \tilde{\mathfrak{Q}}_{\mu,22} : 0 = \varepsilon(\mu(-A_{11}{}^1 + A_{12}{}^2 + 1) - \alpha).$$

We solve these equations for α and $A_{22}{}^2$. The remaining equation can then be solved for μ and we obtain the structure of Assertion 1; $-a + b + 1 \neq 0$ since $\alpha \neq 0$. Since α is unique, we need only test $(x^1)^\alpha \log(x^1)$ and $(x^1)^{\alpha-1}(a_1 x^1 + x^2)$; these fail so $E(\mu, \mathcal{N})$ is 1-dimensional.

Case 2. We assume $A_{22}{}^1 = 0$ but $A_{12}{}^1 \neq 0$. We can rescale to assume $A_{12}{}^1 = 1$ and then fix the gauge by performing a shear to assume $A_{11}{}^1 = 0$. We obtain $\tilde{\mathfrak{Q}}_{\mu,22}: 0 = \mu - A_{22}{}^2 \mu$. Consequently, $A_{22}{}^2 = 1$. We then obtain $\tilde{\mathfrak{Q}}_{\mu,12}: 0 = -\alpha - A_{12}{}^2 \mu$ so $\alpha = -A_{12}{}^2 \mu$. The final equation then becomes $(A_{12}{}^2)^2 \mu(1 + \mu) = 0$. Since $\mu \neq 0$ and $\mu \neq -1$, this forces $A_{12}{}^2 = 0$. This in turn forces $\alpha = 0$, which is false.

Case 3. $A_{22}{}^1 = A_{12}{}^1 = 0$. We obtain $\tilde{\mathfrak{Q}}_{\mu,22} = 0$ and $\tilde{\mathfrak{Q}}_{\mu,12}: 0 = -\frac{1}{2}A_{22}{}^2 \mu$ so $A_{22}{}^2 = 0$ as well. Thus, as we shall establish subsequently in Theorem 14.17, this is linearly equivalent to $\mathcal{N}_1^4(c_2)$, $\mathcal{N}_2^4(\kappa, \theta)$, or $\mathcal{N}_3^4(c_1)$. Since $\mu \neq -1$ and $\rho_s \neq 0$, Lemma 12.32 implies $\dim(E(\mu, \mathcal{N})) \leq 2$. The final equation becomes

$$\tilde{\mathfrak{Q}}_{\mu,11} = \alpha^2 - \alpha(A_{11}{}^1 + 1) + A_{12}{}^2 \mu(-A_{11}{}^1 + A_{12}{}^2 - 1).$$

We can take μ arbitrary and solve this for α. Generically, we will obtain two different values of α; if α is complex we must take real and imaginary part. The discriminant is

$$D = (A_{11}{}^1)^2 + 4A_{11}{}^1 A_{12}{}^2 \mu + 2A_{11}{}^1 - 4(A_{12}{}^2)^2 \mu + 4A_{12}{}^2 \mu + 1.$$

Since $\tilde{\rho} = A_{12}{}^2(A_{11}{}^1 - A_{12}{}^2 + 1)dx^1 \otimes dx^1 \neq 0$, we can solve the equation $D = 0$ for μ to obtain the exceptional value

$$\mu = \frac{(A_{11}{}^1)^2 + 2A_{11}{}^1 + 1}{4A_{12}{}^2(-A_{11}{}^1 + A_{12}{}^2 - 1)}.$$

At this value of μ, we obtain $E(\mu, \mathcal{N}) = \mathrm{span}\{(x^1)^{\frac{1}{2}(1+A_{11}{}^1)}, (x^1)^{\frac{1}{2}(1+A_{11}{}^1)} \log(x^1)\}$. □

It is striking that in the setting of Assertion 1, there is exactly one eigenvalue and the solution space is 1-dimensional. By contrast, in the setting of Assertion 2, the eigenvalue μ can be any real number other than 0 or -1 and the solution space is 2-dimensional.

14.1.6 FLAT TYPE $\mathcal{B}$ GEOMETRIES. Set

$$\mathcal{U}_1(r,s) := \mathcal{N}(1 + rs^2, -s(1 + rs^2), rs, -rs^2, r, -rs), \quad \mathcal{S}_{1,\mathcal{B}}^6 := \text{range}\{\mathcal{U}_1\},$$
$$\mathcal{U}_2(u,v) := \mathcal{N}(u, v, 0, 0, 0, 0), \qquad\qquad\qquad\qquad \mathcal{S}_{2,\mathcal{B}}^6 := \text{range}\{\mathcal{U}_2\},$$
$$\mathcal{U}_3(u,v) := \mathcal{N}(u, v, 0, 1 + u, 0, 0), \qquad\qquad\quad \mathcal{S}_{3,\mathcal{B}}^6 := \text{range}\{\mathcal{U}_3\}.$$

Let $\mathcal{S}_{\mathcal{B}}^6 \subset \mathbb{R}^6$ be the space of flat Type $\mathcal{B}$ models other than the cone point $\mathcal{N}_0^6$. Unlike the Type $\mathcal{A}$ setting described in Theorem 13.34, $\mathcal{S}_{\mathcal{B}}^6$ is not a smooth manifold but consists of the union of three smooth submanifolds of $\mathbb{R}^6$ which intersect transversally along the union of three smooth curves in $\mathbb{R}^6$.

Theorem 14.7 $\mathcal{S}_{\mathcal{B}}^6 = \mathcal{S}_{1,\mathcal{B}}^6 \cup \mathcal{S}_{2,\mathcal{B}}^6 \cup \mathcal{S}_{3,\mathcal{B}}^6$. $\mathcal{S}_{2,\mathcal{B}}^6$ and $\mathcal{S}_{3,\mathcal{B}}^6$ are closed smooth surfaces in $\mathbb{R}^6$ which are diffeomorphic to $\mathbb{R}^2$ and which intersect transversally along the curve $\mathcal{N}(-1, v, 0, 0, 0, 0)$. $\mathcal{S}_{1,\mathcal{B}}^6$ can be completed to a smooth closed surface $\tilde{\mathcal{S}}_{1,\mathcal{B}}^6$ which intersects $\mathcal{S}_{2,\mathcal{B}}^6$ transversally along the curve $\mathcal{N}(1, v, 0, 0, 0, 0)$ and which intersects $\mathcal{S}_{3,\mathcal{B}}^6$ transversally along the curve $\mathcal{N}(0, v, 0, 1, 0, 0)$.

Proof. We use Equation (14.1.a) to compute the Ricci tensor. Let $\Gamma_{ij}{}^k = (x^1)^{-1} A_{ij}{}^k$ where $A_{ij}{}^k \in \mathbb{R}$. Let $\tilde{\rho}_{ij} := (x^1)^2 \rho_{ij}$. A direct computation shows the structures $\mathcal{U}_i(\cdot)$ are flat. We distinguish cases to establish the converse. Because $\tilde{\rho}_{12} - \tilde{\rho}_{21} = A_{12}{}^1 + A_{22}{}^2$, we have that $A_{22}{}^2 = -A_{12}{}^1$.

Case 1. Assume $A_{22}{}^1 \neq 0$. Set $A_{12}{}^1 = rs$, $A_{22}{}^1 = r$, and $A_{22}{}^2 = -rs$ for $r \neq 0$. Then

$$\tilde{\rho}_{22} = -r(1 - A_{11}{}^1 + A_{12}{}^2 + 2rs^2) \quad \text{and} \quad \tilde{\rho}_{21} = -r(A_{11}{}^2 - A_{12}{}^2 s + s).$$

We solve these equations to obtain $A_{11}{}^1 = 1 + A_{12}{}^2 + 2rs^2$ and $A_{11}{}^2 = (A_{12}{}^2 - 1)s$. We have $\tilde{\rho}_{11} = 2(A_{12}{}^2 + rs^2)$. Thus $A_{12}{}^2 = -rs^2$, which gives the parameterization $\mathcal{U}_1$.

Case 2. Suppose $A_{22}{}^1 = 0$. Set $A_{11}{}^1 = u$, $A_{11}{}^2 = v$, and $A_{22}{}^2 = -A_{12}{}^1$ to obtain

$$\tilde{\rho} = \begin{pmatrix} -(A_{12}{}^2)^2 + uA_{12}{}^2 + A_{12}{}^2 - 2A_{12}{}^1 v & A_{12}{}^1(A_{12}{}^2 - 1) \\ A_{12}{}^1(A_{12}{}^2 - 1) & -2(A_{12}{}^1)^2 \end{pmatrix}.$$

This yields $A_{12}{}^1 = 0$ and $A_{12}{}^2(1 + u - A_{12}{}^2) = 0$. If we set $A_{12}{}^2 = 0$, we obtain the parameterization $\mathcal{U}_2$; if we set $A_{12}{}^2 = 1 + u$, we obtain the parameterization $\mathcal{U}_3$. This establishes the first part of the result.

The parameterizations $\mathcal{U}_2$ and $\mathcal{U}_3$ intersect when $u = -1$; the intersection is transversal along the curve $\mathcal{N}(-1, v, 0, 0, 0, 0)$. We wish to extend the parameterization $\mathcal{U}_1$ to study the limiting behavior as $A_{22}{}^1 \to 0$. We distinguish cases.

Case A. Suppose $\lim_{n\to\infty} \mathcal{U}_1(r_n, s_n) \in \text{range}\{\mathcal{U}_2\}$. We have

$$\lim_{n\to\infty}(1 + r_n s_n^2) = u, \quad \lim_{n\to\infty} -s_n(1 + r_n s_n^2) = v, \quad \lim_{n\to\infty} r_n s_n = 0,$$
$$\lim_{n\to\infty} r_n s_n^2 = 0, \qquad\quad \lim_{n\to\infty} r_n = 0, \qquad\qquad \lim_{n\to\infty} r_n s_n = 0.$$

These equations imply $u = 1$, $\lim_{n\to\infty} r_n = 0$, $\lim_{n\to\infty} s_n = -v$. Thus we may simply set $r = 0$ to obtain a transversal intersection along the curve $\mathcal{N}(1, v, 0, 0, 0, 0)$.

Case B. Suppose $\lim_{n\to\infty} \mathcal{U}_1(r_n, s_n) \in \text{range}\{\mathcal{U}_3\}$. We have

$$\lim_{n\to\infty}(1 + r_n s_n^2) = u, \quad \lim_{n\to\infty} -s_n(1 + r_n s_n^2) = v, \quad \lim_{n\to\infty} r_n s_n = 0,$$
$$\lim_{n\to\infty} r_n s_n^2 = 1 + u, \quad \lim_{n\to\infty} r_n = 0, \quad\quad\quad \lim_{n\to\infty} -r_n s_n = 0\,.$$

These equations imply $u = 0$, $\lim_{n\to\infty} r_n = 0$, and $\lim_{n\to\infty} r_n s_n^2 = -1$. We change variables setting $r = -t^2$ and $s = \frac{1}{t} + w$ to express

$$\mathcal{U}_1(-t^2, \tfrac{1}{t} + w) = \mathcal{N}(-2tw - t^2 w^2, 2w + 3tw^2 + t^2 w^3, -t - t^2 w,$$
$$1 + 2tw + t^2 w^2, -t^2, t + t^2 w)\,.$$

We may now safely set $t = 0$ to obtain the intersection with $\text{range}\{\mathcal{U}_3\}$ along the curve $\mathcal{N}(0, 2w, 0, 1, 0, 0)$. $\qquad\square$

14.1.7 THE ALTERNATING RICCI TENSOR. In the Type $\mathcal{B}$ setting, it is possible for the symmetric Ricci tensor to vanish without the geometry being flat; this is not possible in the Type $\mathcal{A}$ setting. For $r \neq 0$ and $u \neq 0$, set:

$$\mathcal{V}_1(r, s, t) \;:=\; \mathcal{N}(s, t, r, 0, 0, r),$$
$$\mathcal{V}_2(u, v, w) \;:=\; \mathcal{N}(1 - 2uw + vw^2, w(1 - uw + vw^2), u - vw, -vw^2, v, u + vw)\,.$$

Theorem 14.8

1. The sets $\text{range}\{\mathcal{V}_i\}$ are smooth 3-dimensional submanifolds of $\mathbb{R}^3$ for $r \neq 0$ and $u \neq 0$. They intersect transversally along a smooth 2-dimensional submanifold.

2. The set of all Type $\mathcal{B}$ structures where $\rho_s = 0$ but $\rho_a \neq 0$ is the union of $\text{range}\{\mathcal{V}_1\}$ and $\text{range}\{\mathcal{V}_2\}$.

3. Let $\mathfrak{V}_1 := \{\mathcal{N}_2^2(c) : c \in \mathbb{R}\}$ and $\mathfrak{V}_2 := \{\mathcal{N}_3^2(0, c, \pm) : c \neq 0\}$. Let G be the $ax + b$ group acting as in Lemma 14.2. The map $(\mathcal{N}, T) \to T^*\mathcal{N}$ is a diffeomorphism from $\mathfrak{V}_i \times G$ to $\mathcal{V}_i$. Thus $\mathcal{V}_i$ can be regarded as a principal $ax + b$ bundle over the curve $\mathfrak{V}_i$.

Proof. It is clear $\text{range}\{\mathcal{V}_1\}$ is a smooth 3-dimensional submanifold. To see $\mathcal{V}_2$ is smooth, we note $u = \frac{1}{2}\{A_{12}{}^1 + A_{22}{}^2\}$ and $v = A_{22}{}^1$. If $v \neq 0$, then $w = \frac{1}{v}(A_{22}{}^2 - u)$, while if $v = 0$, then $w = \frac{1}{2u}(1 - A_{11}{}^1)$. Thus $\mathcal{V}_2$ is 1-1; it is not difficult to verify the Jacobian determinant is non-zero. Thus $\text{range}\{\mathcal{V}_2\}$ is smooth as well. We set $v = 0$ and $u = r$ to see they intersect along the surface $v = 0$, $u = r$, $s = 1 - 2uw$ and $t = w(1 - uw)$. Assertion 1 follows.

A direct computation shows

$$\tilde{\rho}_{\mathcal{V}_1} = r \begin{pmatrix} 0 & 1 \\ -1 & 0 \end{pmatrix} \quad \text{and} \quad \tilde{\rho}_{\mathcal{V}_2} = u \begin{pmatrix} 0 & 1 \\ -1 & 0 \end{pmatrix}\,.$$

So they belong to the set of Type $\mathcal{B}$ models where $\rho_s = 0$ and $\rho_a \neq 0$.

Let $\mathcal{N}$ be a Type $\mathcal{B}$ model with $\rho_s = 0$ and $(\tilde{\rho}_a)_{12} = \frac{1}{2}(A_{12}{}^1 + A_{22}{}^2) \neq 0$. We distinguish cases to prove Assertion 2.

Case 1. Suppose $A_{22}{}^1 = 0$. Set $A_{12}{}^1 = 2r - A_{22}{}^2$ for $r \neq 0$. Setting $\rho_s = 0$ yields

$$(\tilde{\rho}_s)_{11} : \ 0 = A_{12}{}^2 + A_{11}{}^1 A_{12}{}^2 - (A_{12}{}^2)^2 + 2A_{11}{}^2 A_{22}{}^2 - 2A_{11}{}^2 r,$$
$$(\tilde{\rho}_s)_{12} : \ 0 = (1 - A_{12}{}^2)A_{22}{}^2 - r + 2A_{12}{}^2 r,$$
$$(\tilde{\rho}_s)_{22} : \ 0 = 2(3A_{22}{}^2 r - (A_{22}{}^2)^2 - 2r^2).$$

We solve the equation $-2((A_{22}{}^2)^2 - 3A_{22}{}^2 r + 2r^2) = 0$ to obtain $A_{22}{}^2 = r$ or $A_{22}{}^2 = 2r$. Setting $A_{22}{}^2 = 2r$ yields $(\tilde{\rho}_s)_{12}$: $0 = r$ which is false. Consequently, $A_{22}{}^2 = r$. We obtain $(\tilde{\rho}_s)_{12} = A_{12}{}^2 r$ so $A_{12}{}^2 = 0$. Set $A_{11}{}^1 = s$ and $A_{11}{}^2 = t$ to obtain the parameterization $\mathcal{V}_1$.

Case 2. Set $A_{12}{}^1 = 2u - A_{22}{}^2$ and $A_{22}{}^1 = v$ for $u \neq 0$ and $v \neq 0$. We obtain

$$(\tilde{\rho}_s)_{11} : \ 0 = A_{12}{}^2 + A_{11}{}^1 A_{12}{}^2 - (A_{12}{}^2)^2 + 2A_{11}{}^2 A_{22}{}^2 - 2A_{11}{}^2 u,$$
$$(\tilde{\rho}_s)_{12} : \ 0 = (1 - A_{12}{}^2)A_{22}{}^2 - u + 2A_{12}{}^2 u - A_{11}{}^2 v,$$
$$(\tilde{\rho}_s)_{22} : \ 0 = -2(A_{22}{}^2)^2 + 6A_{22}{}^2 u - 4u^2 - (1 - A_{11}{}^1 + A_{12}{}^2)v.$$

Setting $(\tilde{\rho}_s)_{12} = 0$ and $(\tilde{\rho}_s)_{22} = 0$ yields

$$A_{11}{}^2 = \tfrac{1}{v}(A_{22}{}^2 - A_{12}{}^2 A_{22}{}^2 - u + 2A_{12}{}^2 u),$$
$$A_{11}{}^1 = \tfrac{1}{v}(2(A_{22}{}^2)^2 - 6A_{22}{}^2 u + 4u^2 + v + A_{12}{}^2 v).$$

We obtain $(\tilde{\rho}_s)_{11} = \frac{2}{v}((A_{22}{}^2)^2 - 2A_{22}{}^2 u + u^2 + A_{12}{}^2 v)$. This implies

$$A_{12}{}^2 = \tfrac{1}{v}(-(A_{22}{}^2)^2 + 2A_{22}{}^2 u - u^2).$$

Setting $A_{22}{}^2 = vw + u$ yields the parameterization $\mathcal{V}_2$. This parameterization can be extended safely to $v = 0$. Assertion 2 follows.

We apply Lemma 14.2 to compare the parameterizations in order to prove Assertion 3. Again, we distinguish cases.

Case A. Suppose $\mathcal{N} = \mathcal{N}(s, t, r, 0, 0, r)$ is given by the parameterization $\mathcal{V}_1$. Consider a shear $(x^1, x^2) \rightarrow (x^1, ax^2 + bx^1)$ where $a = \frac{1}{r}$ and $b = -\frac{s}{2r}$. By Lemma 14.2 this yields the structure $\mathcal{N}(0, \frac{s^2}{4} + rt, 1, 0, 0, 1)$ which is $\mathcal{N}_2^2(c = \frac{s^2}{4} + rt)$. Thus every element of range$\{\mathcal{V}_1\}$ is linearly equivalent to $\mathcal{N}_2^2(c)$ for some suitably chosen, and unique, c. Conversely, if we apply a shea to $\mathcal{N}_2^2(c) = \mathcal{N}(0, c, 1, 0, 0, 1)$, we obtain $\mathcal{N}(2b, \frac{c-b^2}{a}, a, 0, 0, a)$. To ensure this has the form $\mathcal{N}_2^2(\bar{c})$ for some $\bar{c}$, we need $a = 1$ and $b = 0$; this implies $\bar{c} = c$ as desired. Thus the $ax + b$ group acts without fixed points on $\mathfrak{V}_1$ and the quotient of $\mathfrak{V}_1$ by this action is the curve $\mathcal{N}_2^2(c)$.

Case B. Suppose $\mathcal{N} = \mathcal{V}_2(u, v, w)$. Let $\varepsilon = v$, and $c = u/\varepsilon$. A similar computation taking $a = 1$ and $b = w - c$ yields $\mathcal{N}(1 - c^2\varepsilon, c, 0, -c^2\varepsilon, \varepsilon, 2c\varepsilon)$. We can then rescale x^2 to ensure $\varepsilon = \pm 1$

and obtain $\mathcal{N}_3^2(0, c, \varepsilon)$ for $c \neq 0$. Thus every element of range$\{\mathcal{V}_2(u, v, w)\}$ is represented by some point of the curve $\mathcal{N}_3^2(0, c, \pm)$. Conversely, suppose we apply a shear to $\mathcal{N}_3^2(0, c, \varepsilon)$ and obtain $\mathcal{N}_3^2(0, \bar{c}, \bar{\varepsilon})$. Examining $A_{12}{}^1 = ab\varepsilon$ so $b = 0$. Examining $A_{22}{}^1$ yields $a^2\varepsilon = \bar{\varepsilon}$. Consequently, $\varepsilon = \bar{\varepsilon}$ and $a = \pm 1$. If $a = -1$, we replace c by $-c$ and assume $c > 0$.   □

14.1.8 PARAMETERIZATIONS.   The $ax + b$ group acts on $\mathcal{S}_{\mathcal{B}}^6$ sending (x^1, x^2) to $(x^1, bx^1 + ax^2)$; we permit $a \neq 0$ here to define the notion of linear equivalence; the orbits of this action are the linear equivalence classes in this setting. The analogue of Theorem 13.22 in this case becomes:

Theorem 14.9   *Let $\mathcal{N}$ be a Type $\mathcal{B}$ model.*

1. *$\mathcal{N}$ is flat if and only if $\mathcal{N}$ is linearly equivalent to $\mathcal{N}_i^6(\cdot)$ for some i.*

2. *$\mathcal{N}$ is not flat, but the symmetric Ricci tensor vanishes if and only if $\mathcal{N}$ is linearly equivalent to $\mathcal{N}_3^2(a, c, \pm)$ for $a = 0$ and $c \neq 0$ or to $\mathcal{N}_2^2(c)$.*

3. *$\mathcal{N}$ is strongly projectively flat but not flat if and only if $\mathcal{N}$ is linearly equivalent to $\mathcal{N}_i^4(\cdot), \mathcal{N}_3^3, \mathcal{N}_4^3,$ or $\mathcal{N}_1^2(\cdot)$.*

Proof. We first prove Assertion 1. By Lemma 14.3, the geometries $\mathcal{N}_i^6(\cdot)$ are all flat. Conversely, suppose $\mathcal{N}$ is flat. By Theorem 12.34, $\dim(\mathcal{Q}(\mathcal{N})) = 3$ and $\mathbb{1} \in \mathcal{Q}(\mathcal{N})$. Furthermore, the geometry is determined by $\mathcal{Q}(\mathcal{N})$. We use the parameterization of Theorem 14.7.

Case 1.1. Consider $\mathcal{U}_1(r, s) - \mathcal{N}(1 + rs^2, -s(1 + rs^2), rs, -rs^2, r, -rs)$.

Case 1.1.1. Suppose $r = 0$. We verify $\{\mathbb{1}, x^2 + sx^1, (x^1)^2\} \subset \mathcal{Q}(\mathcal{N})$. We can make a linear change of variables to replace $x^2 + sx^1$ by x^2 and obtain $\mathcal{N}_6^6(1)$.

Case 1.1.2. Suppose $r \neq 0$. We verify $\{\mathbb{1}, sx^1 + x^2, (x^1)^2 + r(sx^1 + x^2)^2\} \subset \mathcal{Q}(\mathcal{N})$. We can make a linear change of variables to replace $|r|^{\frac{1}{2}}(sx^1 + x^2)$ by x^2 and obtain $\mathcal{N}_1^6(\pm)$.

Case 1.2. Consider $\mathcal{U}_2(u, v) = \mathcal{N}(u, v, 0, 0, 0, 0)$.

Case 1.2.1. Suppose $u = -1$. We verify $\{\mathbb{1}, \log(x^1), x^2 + vx^1\} \subset \mathcal{Q}(\mathcal{N})$. We make a linear change of variables to replace $x^2 + vx^1$ by x^2 and obtain $\mathcal{N}_5^6$.

Case 1.2.2. Suppose $u = 0$. We verify $\{\mathbb{1}, x^1, -vx^1 + x^2 + vx^1 \log(x^1)\} \subset \mathcal{Q}(\mathcal{N})$. If $v = 0$, we obtain $\mathcal{N}_0^6$. If $v \neq 0$, we can make a linear change of coordinates to obtain $\mathcal{N}_4^6$.

Case 1.2.3. Suppose $u \notin \{-1, 0\}$. We verify $\{\mathbb{1}, (x^1)^{1+u}, x^2 - \frac{v}{u}x^1\} \subset \mathcal{Q}(\mathcal{N})$. We can make a linear change of variables to replace $x^2 - \frac{v}{u}x^1$ by x^2 and obtain $\mathcal{N}_6^6(c_2)$ for $c_2 = u$. Note that we also obtained $\mathcal{N}_6^6(1)$ previously in Case 1.1.1.

Case 1.3. Consider $\mathcal{U}_3(u, v) = \mathcal{N}(u, v, 0, 1 + u, 0, 0)$.

Case 1.3.1. Suppose $u = -1$. We verify $\{\mathbb{1}, \log(x^1), x^2 + vx^1\} \subset \mathcal{Q}(\mathcal{N})$. We obtain once again $\mathcal{N}_5^6$ since Case 1.3.1 and Case 1.2.1 coincide.

Case 1.3.2. Suppose $u = -2$. We verify $\{\mathbb{1}, \frac{1}{x^1}, \frac{x^2}{x^1} + v \log(x^1)\} \subset \mathcal{Q}(\mathcal{N})$. If $v \neq 0$, we can rescale x^2 to obtain $\mathcal{N}_3^6$. If $v = 0$, we obtain $\mathcal{N}_2^6(-1)$.

Case 1.3.3. Suppose $u \neq -1$ and $u \neq -2$. We verify $\{\mathbb{1}, (x^1)^{1+u}, (x^1)^{1+u}(x^2 + \frac{vx^1}{2+u})\} \subset \mathcal{Q}(\mathcal{N})$. We rescale x^2 to obtain $\mathcal{N}_2^6(1+u)$; the value $u = -2$ was obtained previously in Case 1.3.2.

Assertion 2 follows from Theorem 14.8. Finally, we establish Assertion 3. Suppose $\mathcal{N}$ is strongly projectively flat but not flat. A direct computation shows that any strongly projectively flat surface has both ρ and $\nabla\rho$ totally symmetric; this also follows from Theorem 12.11 of course, but it is not necessary to invoke this result. Set $\tilde{\rho} := (x^1)^2\rho$ and $\bar{\rho}_{ij;k} := (x^1)^3\rho_{ij;k}$ to eliminate denominators. We consider cases.

Case 3.1. Suppose $A_{22}{}^1 = \varepsilon \neq 0$. Let $(\tilde{x}^1, \tilde{x}^2) = (x^1, x^2 + bx^1)$. It follows from Lemma 14.2 that we may choose b so that $A_{12}{}^1 = 0$. Since $\tilde{\rho}_{12} - \tilde{\rho}_{21} = A_{22}{}^2$, $A_{22}{}^2 = 0$. This implies

$$\bar{\rho}_{11;2} - \bar{\rho}_{21;1} = -3A_{11}{}^2\varepsilon \text{ so } A_{11}{}^2 = 0.$$

Finally, $\bar{\rho}_{12;2} - \bar{\rho}_{22;1} = -2\varepsilon(1 - A_{11}{}^1 + 2A_{12}{}^2)$ so $A_{11}{}^1 = 1 + 2A_{12}{}^2$. We set $A_{12}{}^2 = v$ to obtain

$$A_{11}{}^1 = 1 + 2v, A_{11}{}^2 = 0, A_{12}{}^1 = 0, A_{12}{}^2 = v, A_{22}{}^1 = \varepsilon, A_{22}{}^2 = 0,$$
$$\tilde{\rho} = v(2+v)(dx^1)^2 + v\varepsilon(dx^2)^2.$$

We obtain $\{(x^1)^v, x^2(x^1)^v, ((x^1)^2 + \varepsilon(x^2)^2)(x^1)^v\} \subset \mathcal{Q}(\mathcal{N})$. Since $v \neq 0$, we obtain $\mathcal{N}_1^2(v, \pm)$ if $v \neq -1$; if $v = -1$, we obtain $\mathcal{N}_3^3$ or $\mathcal{N}_4^3$.

Case 3.2. Suppose $A_{22}{}^1 = 0$. Assuming ρ and $\nabla\rho$ are symmetric then yields $A_{12}{}^1 = 0$ and $A_{22}{}^2 = 0$ as well. We then have $\tilde{\rho} = (1 + A_{11}{}^1 - A_{12}{}^2)A_{12}{}^2 dx^1 \otimes dx^1$ so $A_{12}{}^2 \neq 0$ and $A_{12}{}^2 \neq 1 + A_{11}{}^1$.

Case 3.2.1. Suppose $A_{12}{}^2 \neq 1 + A_{11}{}^1 - A_{12}{}^2$ and $A_{11}{}^1 \neq 2A_{12}{}^2$. Let $s = -\frac{A_{11}{}^2}{A_{11}{}^1 - 2A_{12}{}^2}$. We then have $\{(x^1)^{A_{12}{}^2}, (x^2 + sx^1)(x^1)^{A_{12}{}^2}, (x^1)^{1+A_{11}{}^1 - A_{12}{}^2}\} \subset \mathcal{Q}(\mathcal{N})$. Change variables to obtain $\mathcal{N}_2^4(\kappa = A_{12}{}^2, \theta = 1 + A_{11}{}^1 - 2A_{12}{}^2 \neq 1)$ with $\kappa \neq 0$, $\theta \neq 0$, $\kappa + \theta \neq 0$.

Case 3.2.2. Suppose $A_{12}{}^2 \neq 1 + A_{11}{}^1 - A_{12}{}^2$ and $A_{11}{}^1 = 2A_{12}{}^2$.

Case 3.2.2.a. If $A_{11}{}^2 = 0$, then $\{(x^1)^{A_{12}{}^2}, (x^1)^{A_{12}{}^2+1}, (x^1)^{A_{12}{}^2}x^2\} \subset \mathcal{Q}(\mathcal{N})$ and we obtain $\mathcal{N}_2^4(\kappa = A_{12}{}^2, \theta = 1)$ with $\kappa \neq 0$ and $\kappa + \theta \neq 0$.

Case 3.2.2.b. If $A_{11}{}^2 \neq 0$, then

$$\{(x^1)^{A_{12}{}^2}, (x^1)^{1+A_{12}{}^2}, (x^2 + A_{11}{}^2x^1\log(x^1))(x^1)^{A_{12}{}^2}\} \subset \mathcal{Q}(\mathcal{N})$$

and we obtain $\mathcal{N}_1^4(c_2 = A_{12}{}^2 \neq 0)$ after rescaling x^2 appropriately.

Case 3.2.3. Suppose finally that $A_{12}{}^2 = 1 + A_{11}{}^1 - A_{12}{}^2$. In this case we have

$$(x^1)^{A_{12}{}^2}\{\mathbb{1}, A_{11}{}^2x^1 + x^2, \log(x^1)\} \subset \mathcal{Q}(\mathcal{N})$$

so after adjusting x^2, we obtain $\mathcal{N}_3^4(c_1 = A_{12}{}^2 \neq 0)$. $\qquad\square$

14.1.9 AFFINE IMMERSIONS AND EMBEDDINGS.

Definition 14.10   Let

$$\Psi_i^6 : \Psi_0^6(x^1, x^2) = (x^1, x^2), \qquad \Psi_1^6(\pm 1)(x^1, x^2)$$
$$= (x^2, (x^1)^2 \pm (x^2)^2),$$

$$\Psi_2^6(c_1)(x^1, x^2) = ((x^1)^{c_1}, (x^1)^{c_1} x^2), \qquad \Psi_3^6(x^1, x^2) = (\tfrac{1}{x^1}, \tfrac{x^2}{x^1} + \log(x^1)),$$
$$\Psi_4^6(x^1, x^2) = (x^1, x^2 + x^1 \log(x^1)) \qquad \Psi_5^6(x^1, x^2) = (\log(x^1), x^2),$$
$$\Psi_6^6(c_2)(x^1, x^2) = ((x^1)^{1+c_2}, x^2).$$

$$\Psi_i^4 : \Psi_1^4(x^1, x^2) = (x^2 + x^1 \log(x^1), \log(x^1)), \quad \Psi_2^4(\kappa, \theta)(x^1, x^2) = (x^2, \theta \log(x^1)),$$
$$\Psi_3^4(c_1)(x^1, x^2) = (x^2, c_1 \log(x^1)).$$

The following result will play the role in the Type $\mathcal{B}$ setting that Lemma 13.12 played in the Type $\mathcal{A}$ setting.

Lemma 14.11

1. $\Psi_i^6(\cdot)$ is an affine embedding of $\mathcal{N}_i^6(\cdot)$ in $\mathcal{M}_0^6$ for any i.
2. Ψ_1^4 is an affine isomorphism from $\mathcal{N}_1^4(c_2)$ to $\mathcal{M}_3^4(c_2)$.
3. $\Psi_2^4(\kappa, \theta)$ is an affine isomorphism from $\mathcal{N}_2^4(\kappa, \theta)$ to $\mathcal{M}_3^4(\tfrac{\kappa}{\theta})$.
4. $\Psi_3^4(c_1)$ is an affine isomorphism from $\mathcal{N}_3^4(c_1)$ to $\mathcal{M}_4^4(0)$.

Proof. By Lemma 13.6 and Theorem 14.9, the geometries in question are all strongly projectively flat. The spaces $\mathcal{Q}(\mathcal{M}_i^j)$ are given in Lemma 13.9 and the spaces $\mathcal{Q}(\mathcal{N}_i^j)$ are given in Lemma 14.4. The diffeomorphisms in question intertwine these solution spaces. The result now follows from Theorem 12.34. □

14.1.10 YAMABE SOLITONS.   We compute the components of the Hessian in this setting:

$$\mathcal{H}_{11}(f) = -(x^1)^{-1}\{A_{11}{}^1 f^{(1,0)} + A_{11}{}^2 f^{(0,1)} - x^1 f^{(2,0)}\},$$
$$\mathcal{H}_{12}(f) = \mathcal{H}_{21}(f) = -(x^1)^{-1}\{A_{12}{}^1 f^{(1,0)} + A_{12}{}^2 f^{(0,1)} - x^1 f^{(1,1)}\},$$
$$\mathcal{H}_{22}(f) = -(x^1)^{-1}\{A_{22}{}^1 f^{(1,0)} + A_{22}{}^2 f^{(0,1)} - x^1 f^{(0,2)}\}.$$

Let $\mathcal{Y}(\mathcal{N}) = E(0, \mathcal{N}) = \ker\{\mathcal{H}\}$ be the space of Yamabe solitons. By Lemma 12.14, we have that $\dim(\mathcal{Y}(\mathcal{N})) \leq 3$. By Lemma 12.40, $\dim(\mathcal{Y}(\mathcal{N})) = 3$ implies $\mathcal{N}$ is flat. In that setting, $\mathcal{Y}(\mathcal{N}) = \mathcal{Q}(\mathcal{N})$ and those structures are treated in Lemma 14.4. Since $\mathbb{1} \in \mathcal{Y}(\mathcal{N})$, we shall be searching for Type $\mathcal{B}$ models where $\dim(\mathcal{Y}(\mathcal{N})) = 2$, i.e., where there is an extra Yamabe soliton.

Lemma 14.12   *Let $\mathcal{N}$ be a Type $\mathcal{B}$ model with $\dim(\mathcal{Y}(\mathcal{N})) = 2$. Then exactly one of the following possibilities pertains.*

1. $A_{12}{}^1 = 0$, $A_{22}{}^1 = 0$, $A_{11}{}^1 \neq -1$, $\mathcal{Y}(\mathcal{N}) = \mathrm{span}\{\mathbb{1}, (x^1)^{1+A_{11}{}^1}\}$, *and*

$$\tilde{\rho} = \begin{pmatrix} -(A_{12}{}^2)^2 + A_{11}{}^1 A_{12}{}^2 + A_{12}{}^2 + A_{11}{}^2 A_{22}{}^2 & A_{22}{}^2 \\ 0 & 0 \end{pmatrix} \neq 0.$$

2. $A_{12}{}^1 = 0$, $A_{22}{}^1 = 0$, $A_{11}{}^1 = -1$, $\mathcal{Y}(\mathcal{N}) = \mathrm{span}\{1, \log(x^1)\}$, and

$$\tilde{\rho} = \begin{pmatrix} A_{11}{}^2 A_{22}{}^2 - (A_{12}{}^2)^2 & A_{22}{}^2 \\ 0 & 0 \end{pmatrix} \neq 0.$$

3. $\mathcal{N}$ is linearly equivalent to $\mathcal{N}(r,0,s,0,t,0)$, $\mathcal{Y}(\mathcal{N}) = \mathrm{span}\{1, x^2\}$, and

$$\tilde{\rho} = \begin{pmatrix} 0 & 0 \\ -s & (r-1)t - s^2 \end{pmatrix} \neq 0.$$

Proof. We apply Lemma 12.53. Let $\mathcal{Y}_\alpha := \ker\{(x^1 \partial_{x^1} + x^2 \partial_{x^2} - \alpha)^2\}$. A priori, α could be complex, although in fact this does not arise. We suppose $\mathcal{Y}_\alpha \neq \{0\}$ and, if $\alpha = 0$, we suppose $\mathcal{Y}_\alpha \neq \mathrm{span}\{1\}$.

Case 1. Suppose $A_{12}{}^1 = 0$, $A_{22}{}^1 = 0$, and $A_{11}{}^1 \neq -1$. We verify that $(x^1)^{1+A_{11}{}^1} \in \mathcal{Y}(\mathcal{N})$. This is the possibility of Assertion 1.

Case 2. Suppose $A_{12}{}^1 = 0$, $A_{22}{}^1 = 0$, and $A_{11}{}^1 = -1$. We verify that $\log(x^1) \in \mathcal{Y}(\mathcal{N})$. This is the possibility of Assertion 2.

Case 3. Suppose $(A_{12}{}^1, A_{22}{}^1) \neq (0,0)$. To clear denominators, we set $\tilde{\mathcal{H}} := (x^1)^{2-\alpha}\mathcal{H}$.

Case 3.1. Suppose $\alpha \neq 0, 1$. By applying $x^1 \partial_{x^1} + x^2 \partial_{x^2} - \alpha$, we may assume that there is no $\log(x^1)$ dependence. Express $f = f_\alpha(x^1) + f_{\alpha-1}(x^1)x^2$. Since

$$f_{\alpha-1} \in E(\alpha-1, \mathcal{N}) \neq E(0, \mathcal{N}),$$

we may assume there is no x^2 dependence. Thus $f = (x^1)^\alpha$. The Hessian equations cannot be satisfied; $\tilde{\mathcal{H}}_{12} : 0 = -\alpha A_{12}{}^1$ and $\tilde{\mathcal{H}}_{22} : 0 = -\alpha A_{22}{}^1$.

Case 3.2. Suppose $\alpha = 0$. Express $f = f_0(x^1) + f_{-1}(x^1)x^2$. Since $\mathcal{Y}_{-1} = \{0\}$, $f_{-1}(x^1) = 0$. Since $1 \in \mathcal{Y}(\mathcal{N})$, we may suppose $f = \log(x^1)$. The Hessian equations cannot be satisfied; $\tilde{\mathcal{H}}_{12}$: $0 = -A_{12}{}^1$ and $\tilde{\mathcal{H}}_{22}$: $0 = -A_{22}{}^1$.

Case 3.3. Suppose $\alpha = 1$. We can eliminate log dependence and suppose $f = a_1 x^1 + a_2 x^2$. If $a_2 = 0$, we obtain $\tilde{\mathcal{H}}_{12}$: $0 = -a_1 A_{12}{}^1$ and $\tilde{\mathcal{H}}_{22}$: $0 = -a_1 A_{22}{}^1$. This is not possible since $(A_{12}{}^1, A_{22}{}^1) \neq (0,0)$. Since $a_2 \neq 0$, $(\hat{x}^1, \hat{x}^2) = (x^1, a_2 x^2 + a_1 x^1)$ defines new coordinates $(\hat{x}^1, \hat{x}^2)$ so that $f(\hat{x}^1, \hat{x}^2) = \hat{x}^2 \in \mathcal{Y}(\mathcal{N})$. The Yamabe soliton equations show that one has $\tilde{\mathcal{H}}_{11}$: $0 = -A_{11}{}^2$, $\tilde{\mathcal{H}}_{12}$: $0 = -A_{12}{}^2$ and $\tilde{\mathcal{H}}_{22}$: $0 = -A_{22}{}^2$. This is the possibility of Assertion 3. $\square$

14.1.11 AFFINE GRADIENT RICCI SOLITONS. Recall that f is said to be an *affine gradient Ricci soliton* if $\mathcal{H}f + \rho_s = 0$. Let $\mathfrak{A}$ be the set of affine gradient Ricci solitons. If $\mathfrak{A}(\mathcal{N})$ is non-empty, then the most general possible affine gradient Ricci soliton can be written in the form $f + \mathcal{Y}(\mathcal{N})$ for any $f \in \mathfrak{A}(\mathcal{N})$. Thus it suffices to exhibit a single element as the space $\mathcal{Y}(\mathcal{N})$ was studied in Lemma 14.12. If $\mathcal{N}$ is flat, we can take $f = \mathbb{1}$, so we eliminate this possibility.

Lemma 14.13 *Let $\mathcal{N}$ be a Type $\mathcal{B}$ model which is not flat. If $\mathcal{N}$ admits an affine gradient Ricci soliton, then one of the following possibilities pertains.*

1. $A_{12}{}^1 = 0$, $A_{22}{}^1 = 0$, $A_{22}{}^2 = 0$, $A_{11}{}^1 \neq -1$; $\frac{1+A_{11}{}^1-A_{12}{}^2}{1+A_{11}{}^1} A_{12}{}^2 \log(x^1) \in \mathfrak{A}(\mathcal{N})$.
2. $A_{12}{}^1 = 0$, $A_{22}{}^1 = 0$, $A_{22}{}^2 = 0$, $A_{11}{}^1 = -1$; $\frac{1}{2}(A_{12}{}^2)^2 (\log(x^1))^2 \in \mathfrak{A}(\mathcal{N})$.
3. $\mathcal{N}$ *is linearly equivalent to* $\mathcal{N}_3^2(a,c,\pm)$; $a \log(x^1) \in \mathfrak{A}(\mathcal{N})$.
4. $\mathcal{N}$ *is linearly equivalent to* $\mathcal{N}_2^2(c)$; $0 \in \mathfrak{A}(\mathcal{N})$.

We shall show in Theorem 14.17 that the geometries discussed in Assertion 1 are also Type $\mathcal{A}$.

Proof. A direct computation shows that the structures of Assertions 1–4 admit the gradient Ricci solitons listed. To complete the proof of the lemma, we must establish the converse implication. To clear denominators, we normalize the soliton equations defining

$$\tilde{S} := (x^1)^2 S = (x^1)^2 \{\mathcal{H}f + \rho_s\}.$$

Assume that $f \in \mathfrak{A}(\mathcal{N})$ and $\mathcal{N}$ is not flat. If $\rho_s = 0$, then $\rho_a \neq 0$ and Theorem 14.9 shows either $\mathcal{N}$ is linearly equivalent to $\mathcal{N}_2^2(c)$ or to $\mathcal{N}_3^2(0,c,\pm)$. Thus we shall suppose that $\rho_s \neq 0$ henceforth and, consequently, that $\mathfrak{A}(\mathcal{N})$ is disjoint from $\mathcal{Y}(\mathcal{N})$. If X is an affine Killing vector field, then $X(f) \in \mathcal{Y}(\mathcal{N})$. We will use this observation in our analysis.

Assertions 1–2. Suppose $A_{12}{}^1 = 0$ and $A_{22}{}^1 = 0$. By Lemma 14.12, Yamabe solitons are independent of x^2. Since $\partial_{x^2} f \in \mathcal{Y}(\mathcal{N})$, $f = f_0(x^1) + f_1(x^1)x^2$. We have

$$\tilde{S}_{22} : \ 0 = -A_{22}{}^2 x^1 f_1(x^1).$$

If $f_1 \neq 0$, then $A_{22}{}^2 = 0$. If $f_1 = 0$, we obtain $\tilde{S}_{12}$: $0 = \frac{1}{2}A_{22}{}^2$ so $A_{22}{}^2 = 0$. Thus in either event, $A_{22}{}^2 = 0$ and the possibilities of Assertions 1 and 2 hold.

Assertions 3–4. We apply Lemma 14.12. Suppose $(A_{12}{}^1, A_{22}{}^1) \neq (0,0)$. Then either

$$\mathcal{Y}(\mathcal{N}) = \text{span}\{\mathbb{1}, x^2\} \text{ or } \mathcal{Y}(\mathcal{N}) = \text{span}\{\mathbb{1}\}.$$

Suppose $\mathcal{Y}(\mathcal{N}) = \text{span}\{\mathbb{1}, x^2\}$. Since $\partial_{x^2} f \in \mathcal{Y}(\mathcal{N})$, $f = f_0(x^1) + x^2(a_0 + a_1 x^2)$. We apply the affine Killing vector field $x^1 \partial_{x^1} + x^2 \partial_{x^2}$ to see

$$x^1 f_0'(x^1) + x^2(a_0 + 2a_1 x^2) = (x^1 \partial_{x^1} + x^2 \partial_{x^2}) f \in \mathcal{Y}(\mathcal{N}).$$

Consequently, $a_1 = 0$. Since $x^2 \in \mathcal{Y}(\mathcal{N})$, we can delete the $a_0 x^2$ term and assume $f = f_0(x^1)$. Since $x^1 f_0' = a_2$, we have $f_0 = a_2 \log(x^1)$ since we may remove the constant term. On the other hand, if $\mathcal{Y}(\mathcal{N}) = \text{span}\{\mathbb{1}\}$, then $f = f_0(x^1) + a_0 x^2$. Because $x^1 f_0'(x^1) + a_0 x^2 \in \mathcal{Y}(\mathcal{N})$, we may assume that $a_0 = 0$. Again we have $x^1 f_0' = a_2$. Consequently, we shall suppose that $f = a \log(x^1)$.

Case 1. Suppose $A_{22}{}^1 \neq 0$. We can normalize $A_{22}{}^1 = \varepsilon = \pm 1$. We use Lemma 14.2 to choose a shear so that we may assume $A_{12}{}^1 = 0$. We set $A_{11}{}^2 = c$. We have

$$\tilde{S}_{12} : 0 = \tfrac{1}{2}(A_{22}{}^2 - 2c\varepsilon) \quad \text{and} \quad \tilde{S}_{22} : 0 = \varepsilon(A_{11}{}^1 - A_{12}{}^2 - 1 - a).$$

This yields $A_{22}{}^2 = 2c\varepsilon$ and $A_{11}{}^1 = 1 + a + A_{12}{}^2$. The remaining soliton equation is then

$$\tilde{S}_{11} : 0 = -2a - a^2 + 2A_{12}{}^2 + 2c^2\varepsilon.$$

We use this equation to determine $A_{12}{}^2$ and establish Assertion 3 by computing:

$$\mathcal{N} = \mathcal{N}(\tfrac{1}{2}(2 + 4a + a^2 - 2c^2\varepsilon), c, 0, a + \tfrac{1}{2}a^2 - c^2\varepsilon, \varepsilon, 2c\varepsilon) = \mathcal{N}_3^2(a, c, \pm).$$

Case 2. Suppose $A_{22}{}^1 = 0$ but $A_{12}{}^1 \neq 0$. Rescale to set $A_{12}{}^1 = 1$. Set $c := A_{11}{}^2$ and obtain

$$\tilde{S}_{12} : 0 = \tfrac{1}{2}(-1 - 2a + 2A_{12}{}^2 + A_{22}{}^2) \quad \text{and} \quad \tilde{S}_{22} : 0 = -1 + A_{22}{}^2.$$

We obtain $A_{22}{}^2 = 1$ and $A_{12}{}^2 = a$; $\tilde{S}_{11}: 0 = -a^2$ yields $a = 0$; the Ricci tensor is alternating. By Theorem 14.9, $A_{22}{}^1 = 0$ and $\mathcal{N} = \mathcal{N}_2^2(c)$; Assertion 4 follows.   □

14.2   TYPE $\mathcal{B}$: AFFINE KILLING VECTOR FIELDS

In this section, we present some results related to the Lie algebra of affine Killing vector fields. In Section 14.2.1, we study the space of affine Killing vector fields for flat geometries, and in Section 14.2.2, we perform a similar analysis for the non-flat geometries. In Section 14.2.3, we classify the models where $\dim(\mathfrak{K}) > 2$. We use these results in Section 14.2.4 to classify the non-flat models which are both Type $\mathcal{A}$ and Type $\mathcal{B}$ models. In Section 14.2.5, we determine which Type $\mathcal{B}$ models are affine Killing complete. In Section 14.2.6, we examine invariant and parallel tensors of Type $(1,1)$.

14.2.1 AFFINE KILLING VECTOR FIELDS FOR FLAT GEOMETRIES.

Lemma 14.14   *The geometries $\mathcal{N}_i^6(\cdot)$ are all flat. Consequently, $\dim(\mathfrak{K}(\mathcal{N}_i^6(\cdot))) = 6$. Let a_i be arbitrary real constants.*

1. $\mathfrak{K}(\mathcal{N}_0^6) = \mathrm{span}\left\{(a_1 + a_2 x^1 + a_3 x^2)\partial_{x^1} + (a_4 + a_5 x^1 + a_6 x^2)\partial_{x^2}\right\}$.

2. $\mathfrak{K}(\mathcal{N}_1^6(\pm)) = \mathrm{span}\{Z_1 \partial_{x^1} + Z_2 \partial_{x^2}\}$ *where*

$$Z_1 = a_1 x^1 + \tfrac{a_2}{x^1} + \tfrac{a_3 x^2}{x^1} + \tfrac{a_4((x^1)^2 \pm (x^2)^2)}{x^1} - \tfrac{a_6 x^2(\pm(x^1)^2 + (x^2)^2)}{x^1},$$
$$Z_2 = a_1 x^2 + a_5 + a_6\left((x^1)^2 \pm (x^2)^2\right).$$

3. $\mathfrak{K}(\mathcal{N}_2^6(c_1)) = \mathrm{span}\{Z_1 \partial_{x^1} + Z_2 \partial_{x^2}\}$ *where*

$$Z_1 = x^1\left(a_1 + a_2 x^2 + a_3(x^1)^{-c_1}\right),$$
$$Z_2 = -a_2 c_1(x^2)^2 - a_3 c_1 x^2(x^1)^{-c_1} + a_4 x^2 + a_5 + a_6(x^1)^{-c_1}.$$

4. $\mathfrak{K}(\mathcal{N}_3^6) = \text{span}\,\{Z_1\partial_{x^1} + Z_2\partial_{x^2}\}$ *where for* $\tau(x^1, x^2) := x^2 + x^1\log(x^1)$ *we have*

$$Z_1 = a_1 x^1 - a_2(x^1)^2 - a_3 x^1 \tau(x^1, x^2),$$
$$Z_2 = a_1 x^2 + a_2\left((x^1)^2 - x^1 x^2\right) + a_3(x^1 - x^2)\tau(x^1, x^2) + a_4 + a_5 x^1$$
$$+ a_6 \tau(x^1, x^2).$$

5. $\mathfrak{K}(\mathcal{N}_4^6) = \text{span}\,\{Z_1\partial_{x^1} + Z_2\partial_{x^2}\}$ *where we have for* $\tau(x^1, x^2) := x^2 + x^1\log(x^1)$ *and for* $\tau_1(x^1) := \log(x^1) + 1$ *that*

$$Z_1 = a_1 x^1 + a_2\tau(x^1, x^2) + a_3,$$
$$Z_2 = a_1 x^2 - a_2\tau_1(x^1)\tau(x^1, x^2) - a_3\tau_1(x^1) + a_4 + a_5 x^1 + a_6\tau(x^1, x^2).$$

6. $\mathfrak{K}(\mathcal{N}_5^6) = \text{span}\,\left\{\left(x^1(a_1 + a_2 x^2 + a_3\log(x^1))\right)\partial_{x^1} + \left(a_4 + a_5 x^2 + a_6\log(x^1)\right)\partial_{x^2}\right\}$.

7. $\mathfrak{K}(\mathcal{N}_6^6(c_2)) = \text{span}\,\{Z_1\partial_{x^1} + Z_2\partial_{x^2}\}$ *where*

$$Z_1 = a_1 x^2(x^1)^{-c_2} + a_2(x^1)^{-c_2} + a_3 x^1,$$
$$Z_2 = a_4 + a_5 x^2 + a_6(x^1)^{c_2+1}.$$

Proof. Let $\mathcal{N}$ be a Type $\mathcal{B}$ geometry. By Lemma 12.53, the components of an affine Killing vector field are sums of elements of the form $(x^1)^\alpha p(x^2, \log(x^1))$ where p is polynomial. This fact informed our investigation. Suppose $\mathcal{N}$ is flat. Let $\mathcal{Q}(\mathcal{N}) = \text{span}\{1, \phi_1, \phi_2\}$. By Lemma 12.32, $\Phi := (\phi_1, \phi_2)$ defines an immersion. Since $\Phi^*\mathcal{Q}(\mathcal{M}_0^6) = \mathcal{Q}(\mathcal{N})$, Φ is an affine immersion by Theorem 12.34. Consequently, the affine Killing vector fields of $\mathcal{N}$ are the pull-back of the affine Killing vector fields of $\mathcal{M}_0^6$. Since the elements of $\mathfrak{K}(\mathcal{M}_0^6)$ take the form $a_i\,\partial_{x^i} + a_i^j\,x^i\partial_{x^j}$, this provides an algorithm for determining $\mathfrak{K}(\mathcal{N})$. We used the computation of $\mathcal{Q}(\mathcal{N})$ given in Lemma 14.4 to inform our investigation. One can now check directly that the vector fields given in the lemma are in fact affine Killing vector fields; they span for dimensional reasons. We emphasize, this was not a routine mathematica calculation; it was informed by the theory noted above. □

14.2.2 AFFINE KILLING VECTOR FIELDS FOR NON-FLAT MODELS.

Lemma 14.15 *For the non-flat geometries of Definition 14.1 we have that:*

1. $\mathfrak{K}(\mathcal{N}_1^4(c_2)) = \text{span}\,\left\{\left(a_1 x^1 + a_2 x^1\right)\partial_{x^1} + \left(a_1 x^2 - a_2 x^1\log(x^1) + a_3 + a_4 x^1\right)\partial_{x^2}\right\}$.

2. $\mathfrak{K}(\mathcal{N}_2^4(\kappa, \theta)) = \text{span}\,\left\{a_1 x^1\partial_{x^1} + \left(a_2 x^2 + a_3 + a_4(x^1)^\theta\right)\partial_{x^2}\right\}$.

3. $\mathfrak{K}(\mathcal{N}_3^4(c_1)) = \text{span}\,\left\{a_1 x^1\partial_{x^1} + \left(a_2 x^2 + a_3 + a_4\log(x^1)\right)\partial_{x^2}\right\}$.

4. $\mathfrak{K}(\mathcal{N}_1^3(\pm)) = \mathfrak{K}(\mathcal{N}_2^3(c)) = \text{span}\,\left\{\left(a_1 x^1 + 2a_2 x^1 x^2\right)\partial_{x^1} + \left(a_1 x^2 + a_2(x^2)^2 + a_3\right)\partial_{x^2}\right\}$.

5. $\mathfrak{K}(\mathcal{N}_3^3) = \text{span}\,\left\{\left(a_1 x^1 + 2a_2 x^1 x^2\right)\partial_{x^1} + \left(a_1 x^2 + a_2((x^2)^2 + (x^1)^2) + a_3\right)\partial_{x^2}\right\}$.

6. $\mathfrak{K}(\mathcal{N}_4^3) = \text{span}\,\left\{\left(a_1 x^1 + 2a_2 x^1 x^2\right)\partial_{x^1} + \left(a_1 x^2 + a_2((x^2)^2 - (x^1)^2) + a_3\right)\partial_{x^2}\right\}$.

7. $\mathfrak{K}(\mathcal{N}_i^2(\cdot)) = \text{span}\,\left\{a_1 x^1\partial_{x^1} + \left(a_1 x^2 + a_2\right)\partial_{x^2}\right\}$ *for all* $1 \le i \le 3$ *excepting the case* $\mathcal{N}_3^2(0, \frac{3}{\sqrt{2}}, +)$, *which is equivalent to* $\mathcal{N}_2^3\left(\frac{1}{2}\right)$.

Proof. We perform a direct computation to see the vector fields given in the lemma are actually affine Killing vector fields. We use Lemma 14.11 to verify that the structures $\mathcal{N}_i^4(\cdot)$ are isomorphic to Type $\mathcal{A}$ manifolds $\mathcal{M}_j^4(\cdot)$. Consequently, $\dim(\mathfrak{K}(\mathcal{N}_i^4(\cdot))) = 4$, which proves Assertions 1–3. Using the ansatz of Lemma 12.53, one verifies there are no additional affine Killing vector fields to establish the remaining assertions.   □

14.2.3 CLASSIFICATION OF TYPE $\mathcal{B}$ MODELS WITH DIM($\mathfrak{K}$)> 2.

Theorem 14.16   *Let $\mathcal{N}$ be a Type $\mathcal{B}$ model.*

1. $\dim(\mathfrak{K}(\mathcal{N})) \in \{2, 3, 4, 6\}$.

2. $\dim(\mathfrak{K}(\mathcal{N})) = 3$ *if and only if $\mathcal{N}$ is linearly equivalent to $\mathcal{N}_i^3(\cdot)$ for some i.*

3. *The following assertions are equivalent.*

    (a) $\dim(\mathfrak{K}(\mathcal{N})) = 4$.

    (b) $\mathcal{N}$ is linearly equivalent to $\mathcal{N}_i^4(\cdot)$ for some i;

    (c) $\mathcal{N}$ is also Type $\mathcal{A}$ and not flat.

4. *The following assertions are equivalent.*

    (a) $\dim(\mathfrak{K}(\mathcal{N})) = 6$.

    (b) $\mathcal{N}$ is linearly equivalent to $\mathcal{N}_i^6(\cdot)$ for some i.

    (c) $\mathcal{N}$ is flat.

Proof. We distinguish cases. Assertion 4 follows from Theorem 14.9, so we shall assume $\mathcal{N}$ is not flat henceforth. Assertions 1–3 were first established by Brozos-Vázquez, García-Río, and Gilkey [7]. The variable $A_{22}{}^1$ plays a central role in our analysis as it is unchanged by any shear. Our discussion is informed by the the corresponding discussion which was used to prove Lemma 12.53. We will examine the generalized eigenspaces $E(\alpha)$ of the adjoint map $\mathrm{ad}(x^1\partial_{x^1} + x^2\partial_{x^2})$; we change notation slightly and shift the eigenvalue to define

$$E(\alpha) := \{X \in \mathfrak{K}_{\mathbb{C}}(\mathcal{N}) : (\mathrm{ad}(x^1\partial_{x^1} + x^2\partial_{x^2}) - (\alpha - 1))^6 X = 0\};$$

thus $E(\alpha)$ is generated by elements of the form $(x^1)^\beta (x^2)^{\alpha-\beta} \log(x^1)^j$, which are homogeneous of degree α in $\{x^1, x^2\}$ jointly; the variable $\log(x^1)$ contributes Jordan normal form but is homogeneous of degree 0. Clearly $\mathrm{ad}(\partial_{x^2}) : E(\alpha) \to E(\alpha - 1)$. Assume $\dim(\mathfrak{K}(\mathcal{N})) > 2$ so there is an element $X \in E(\alpha)$ for some α which is neither $\partial_{x^2} \in E(0)$ nor $x^1\partial_{x^1} + x^2\partial_{x^2} \in E(1)$. Let $K_{ij}{}^k$ be the affine Killing equations of Equation (12.1.d).

Case 1. Suppose $A_{22}{}^1 \neq 0$. Except when considering $E(2)$, we will apply Lemma 14.2 to see there exists a unique shear $(x^1, x^2) \to (x^1, x^2 + bx^1)$ so that $A_{12}{}^1 = 0$. We then rescale x^2 to ensure $A_{22}{}^1 = \varepsilon = \pm 1$. This fixes the gauge in these cases; when considering $E(2)$, a different gauge normalization will be convenient.

Case 1.1. Suppose $\Re(\alpha) < 0$ and $\alpha \neq -1$. Choose α so $\Re(\alpha)$ is minimal. As $\mathrm{ad}(\partial_{x^2})X$ belongs to $E(\alpha - 1)$ and since $\Re(\alpha)$ is minimal, $\mathrm{ad}(\partial_{x^2})X = 0$ so there is no x^2 dependence in X. By applying $\mathrm{ad}(x^1\partial_{x^1} + x^2\partial_{x^2}) - \alpha$, we may also assume there is no $\log(x^1)$ dependence. Consequently, $X = (x^1)^\alpha\{a_1\partial_{x^1} + a_2\partial_{x^2}\}$. We show $X = 0$ and thus this case does not happen by considering

$$K_{12}{}^1 : 0 = (x^1)^{\alpha-2}a_2\alpha\varepsilon \quad \text{and} \quad K_{22}{}^1 : 0 = -(x^1)^{\alpha-2}a_1(1+\alpha)\varepsilon . \tag{14.2.a}$$

Case 1.2. Suppose $\alpha = -1$. As in Case 1.1, we may assume $X = (x^1)^{-1}(a_1\partial_{x^1} + a_2\partial_{x^2})$; Equation (14.2.a) then implies $a_2 = 0$ so $X = (x^1)^{-1}\partial_{x^1}$. We obtain a flat geometry which is false:

$$\begin{aligned}
K_{11}{}^1 : 0 &= 2(x^1)^{-3}(1 - A_{11}{}^1), \quad K_{11}{}^2 : 0 = -3(x^1)^{-3}A_{11}{}^2, \\
K_{12}{}^2 : 0 &= -2(x^1)^{-3}A_{12}{}^2, \quad K_{22}{}^2 : 0 = -(x^1)^{-3}A_{22}{}^2, \quad \rho = 0 .
\end{aligned}$$

Case 1.3. Suppose $\alpha = 0$. Any non-trivial x^2 dependence would give rise to an element of $E(-1)$, which has been dealt with in Case 1.2. If there is a $\log(x^1)^k$ dependence where $k > 1$, we may apply $\mathrm{ad}(x^1\partial_{x^1} + x^2\partial_{x^2})^{k-1}$ to assume $k = 1$. We may therefore assume that

$$X = (a_1 + a_2\log(x^1))\partial_{x^1} + (b_1 + b_2\log(x^1))\partial_{x^2}.$$

Since $\partial_{x^2} \in \mathfrak{K}(\mathcal{N})$, we may assume without loss of generality that $b_1 = 0$ since we want an additional vector field. The following equations imply $a_1 = 0, a_2 = 0$, and $b_2 = 0$ which is false:

$$\begin{aligned}
K_{22}{}^1 : 0 &= -\varepsilon(x^1)^{-2}(a_1 + a_2 + a_2\log(x^1)), \\
K_{22}{}^2 : 0 &= -(x^1)^{-2}(a_1 A_{22}{}^2 + b_2\varepsilon + a_2 A_{22}{}^2\log(x^1)) .
\end{aligned}$$

Case 1.4. Suppose $\alpha = 1$. Any non-trivial x^2 dependence divided by a power of x^1 would give rise to an element of $E(\alpha)$ for $\alpha < 0$ which has already been eliminated. Similarly, any $\log(x^1)^2$ dependence can be eliminated. So we assume

$$\begin{aligned}
X = {} & (a_1 x^1 + a_2 x^2 + a_3 x^1\log(x^1) + a_4 x^2\log(x^1))\partial_{x^1} \\
& + (b_1 x^1 + b_2 x^2 + b_3 x^1\log(x^1) + b_4 x^2\log(x^1))\partial_{x^2} .
\end{aligned}$$

Since $x^1\partial_{x^1} + x^2\partial_{x^2}$ belongs to $\mathfrak{K}(\mathcal{N})$, we may assume $b_2 = 0$ to eliminate this term. We have

$$\mathrm{ad}(\partial_{x^2})X = (a_2 + a_4\log(x^1))\partial_{x^1} + (b_2 + b_4\log(x^1))\partial_{x^2} .$$

By Case 1.3, we have that $a_2 = a_4 = b_4 = 0$. We show that $a_1 = a_3 = b_1 = b_3 = 0$ and, consequently, there are no unexpected terms which are homogeneous of degree 1 by computing

$$\begin{aligned}
X &= (a_1 x^1 + a_3 x^1\log(x^1))\partial_{x^1} + (b_1 x^1 + b_3 x^1\log(x^1))\partial_{x^2}, \\
K_{12}{}^1 : 0 &= (x^1)^{-1}\varepsilon(b_1 + b_3 + b_3\log(x^1)), \\
K_{22}{}^1 : 0 &= -(x^1)^{-1}\varepsilon(2a_1 + a_3 + 2a_3\log(x^1)) .
\end{aligned}$$

Case 1.5. Suppose $\alpha = 2$. We recall that we use a different gauge normalization in this case; we will obtain all the $\mathcal{N}_i^3(\cdot)$ models except $\mathcal{N}_2^3(0)$. The arguments given previously show we may suppose

$$X = (a_{11}x^1x^1 + 2a_{12}x^1x^2 + a_{22}x^2x^2)\partial_{x^1} + (b_{11}x^1x^1 + 2b_{12}x^1x^2 + b_{22}x^2x^2)\partial_{x^2}.$$

We have $\mathrm{ad}(\partial_{x^2})^2 X = a_{22}\partial_{x^1} + b_{22}\partial_{x^2}$; $a_{22} = 0$ by Case 1.1. We apply Case 1.4 to see $b_{12} = 0$ and $b_{22} = a_{12}$ since $\mathrm{ad}(\partial_{x^2})X = 2a_{12}x^1\partial_{x^1} + (2b_{12}x^1 + 2b_{22}x^2)\partial_{x^2}$. Consequently,

$$X = (a_{11}(x^1)^2 + 2a_{12}x^1x^2)\partial_{x^1} + (b_{11}(x^1)^2 + a_{12}(x^2)^2)\partial_{x^2}.$$

We set $A_{22}^{\ 1} = \varepsilon \neq 0$, but we do not normalize $A_{12}^{\ 2}$. If $a_{12} = 0$, then $K_{22}^{\ 1}: 0 = -3a_{11}\varepsilon$ and $K_{22}^{\ 2} = -a_{11}A_{22}^{\ 2} - 2b_{11}\varepsilon$. This would imply $X = 0$, which is false. Therefore $a_{12} \neq 0$, so we may assume $a_{12} = 1$. Set $\tilde{x}^2 := x^2 + \frac{1}{2}a_{11}x^1$ to ensure $X = 2x^1\tilde{x}^2\partial_{x^1} + \star\partial_{x^2}$. This fixes the gauge up to a possible rescaling of x^2. We have

$$X = 2x^1x^2\partial_{x^1} + (b_{11}(x^1)^2 + (x^2)^2)\partial_{x^2},$$
$$K_{12}^{\ 1}: 0 = 2(A_{11}^{\ 1} - A_{12}^{\ 2} + b_{11}\varepsilon + 1),$$
$$K_{22}^{\ 1}: 0 = 4A_{12}^{\ 1} - 2A_{22}^{\ 2}.$$

This determines b_{11} and shows $A_{22}^{\ 2} = 2A_{12}^{\ 1}$. We determine $A_{11}^{\ 2}$ from the affine Killing equation $K_{12}^{\ 2}: 0 = 2(A_{11}^{\ 2} + \frac{1}{\varepsilon}A_{12}^{\ 1}(-1 - A_{11}^{\ 1} + A_{12}^{\ 2}))$ and then determine $A_{11}^{\ 1}$ from the affine Killing equation $K_{22}^{\ 2}: 0 = 2(2 + A_{11}^{\ 1} + A_{12}^{\ 2})$. The remaining two affine Killing equations become

$$K_{11}^{\ 1}: 0 = \tfrac{6}{\varepsilon}A_{12}^{\ 1}(1 + 2A_{12}^{\ 2}) \quad \text{and} \quad K_{11}^{\ 2}: 0 = \tfrac{6}{\varepsilon}(1 + 3A_{12}^{\ 2} + 2(A_{12}^{\ 2})^2).$$

We solve the equation $1 + 3A_{12}^{\ 2} + 2(A_{12}^{\ 2})^2 = 0$ to see either $A_{12}^{\ 2} = -\frac{1}{2}$ or $A_{12}^{\ 2} = -1$.

Case 1.5.1. $A_{12}^{\ 2} = -\frac{1}{2}$. We have $X = 2x^1x^2\partial_{x^1} + (x^2)^2\partial_{x^2}$ satisfies the affine Killing equations with $\mathcal{N} = \mathcal{N}(-\frac{3}{2}, 0, A_{12}^{\ 1}, -\frac{1}{2}, \varepsilon, 2A_{12}^{\ 1})$. If $A_{12}^{\ 1} = 0$, we obtain $\mathcal{N}(-\frac{3}{2}, 0, 0, -\frac{1}{2}, \varepsilon, 0)$. We can rescale x^2 to ensure $\varepsilon = \pm\frac{1}{2}$ and obtain $\mathcal{N}_1^3(\pm)$. If $A_{12}^{\ 1} \neq 0$, we can rescale x^2 to ensure $A_{12}^{\ 1} = 1$ and obtain $\mathcal{N}(-\frac{3}{2}, 0, 1, -\frac{1}{2}, \varepsilon, 2)$. This is $\mathcal{N}_2^3(c)$ for $c = \varepsilon \neq 0$.

Case 1.5.2. $A_{12}^{\ 2} = -1$. $K_{11}^{\ 1}: 0 = -\frac{6}{\varepsilon}A_{12}^{\ 1}$. Set $A_{12}^{\ 1} = 0$ so $\mathcal{N} = \mathcal{N}(-1, 0, 0, -1, \varepsilon, 0)$ and $X = 2x^1x^2\partial_{x^1} + ((x^2)^2 - \frac{1}{\varepsilon}(x^1)^2)\partial_{x^2}$. Rescale $\varepsilon = \pm 1$ to obtain either $\mathcal{N}_3^3$ or $\mathcal{N}_4^3$.

Case 1.6. For $\alpha \geq 0$ we have dealt with the eigenspaces $E(0)$, $E(1)$, and $E(2)$. If $\alpha \notin \{0, 1, 2\}$, we could apply $\mathrm{ad}(\partial_{x^2})$ to eliminate any x^2 terms and $\mathrm{ad}(x^1\partial_{x^1} + x^2\partial_{x^2})$ to eliminate any $\log(x^1)$ terms to ensure $X = (x^1)^\alpha(a_1\partial_{x^1} + a_2\partial_{x^2})$; this is eliminated by Equation (14.2.a).

Case 2. Suppose $A_{22}^{\ 1} = 0$. We will obtain $\mathcal{N}_2^3(0)$ and $\mathcal{N}_i^4(\cdot)$.

Case 2.1. Suppose $X = (x^1)^\alpha(a_1\partial_{x^1} + a_2\partial_{x^2})$ where $\alpha \neq 0, 1$; if $\Re(\alpha) < 0$, the argument of Case 1.1 permits us to assume that X has this form whereas for other values of α, it is a separate assumption.

Case 2.1.1. If $a_1 \neq 0$, then $K_{12}{}^1 \colon 0 = -a_1 A_{12}{}^1 (x^1)^{\alpha-2}$ and $K_{22}{}^2 \colon 0 = -a_1 A_{22}{}^2 (x^1)^{\alpha-2}$. Thus $A_{12}{}^1 = A_{22}{}^2 = 0$. We obtain $K_{12}{}^2 \colon 0 = a_1 A_{12}{}^2 (x^1)^{\alpha-2} (\alpha - 1)$, so $A_{12}{}^2 = 0$ and $\rho = 0$, which is false.

Case 2.1.2. Suppose $a_1 = 0$. Set $X = (x^1)^{\alpha} \partial_{x^2}$. We have that

$$K_{11}{}^1 \colon \ 0 = 2(x^1)^{\alpha-2} \alpha A_{12}{}^1, \qquad\qquad K_{11}{}^2 \colon \ 0 = (x^1)^{\alpha-2} \alpha (-1 - A_{11}{}^1 + 2 A_{12}{}^2 + \alpha),$$
$$K_{12}{}^2 \colon \ 0 = (x^1)^{\alpha-2} \alpha (A_{22}{}^2 - A_{12}{}^1).$$

We set $\alpha = 1 + A_{11}{}^1 - 2 A_{12}{}^2$, $A_{12}{}^1 = 0$, and $A_{22}{}^2 = 0$. Since $\alpha \neq 1$, $A_{11}{}^1 - 2 A_{12}{}^2 \neq 0$ and we can apply Lemma 14.2 to perform a shear to ensure $A_{11}{}^2 = 0$. Set $A_{11}{}^1 := 2\kappa + \theta - 1$ and $A_{12}{}^2 = \kappa$. Then $(x^1)^{\theta} \partial_{x^2} \in \mathfrak{K}(\mathcal{N})$ for $\mathcal{N}(\theta + 2\kappa - 1, 0, 0, \kappa, 0, 0) = \mathcal{N}_2^4(\kappa, \theta)$. Because we have $\tilde{\rho} = \kappa(\kappa + \theta) dx^1 \otimes dx^1 \neq 0$, we obtain $\kappa(\theta + \kappa) \neq 0$. Because $\alpha \neq 0$, $\theta \neq 0$.

Case 2.2. Suppose $\alpha = 0$. We first assume there are no log terms and, since ∂_{x^2} is an affine Killing vector field, examine $X = \partial_{x^1}$. The affine Killing equations yield $\Gamma = 0$, which is a flat geometry. Thus ∂_{x^1} is not an affine Killing vector field. Suppose $X = \log(x^1) \partial_{x^2}$.

$$K_{11}{}^1 \colon \ 0 = (x^1)^{-2} 2 A_{12}{}^1, \qquad\qquad K_{11}{}^2 \colon \ 0 = -(x^1)^{-2}(1 + A_{11}{}^1 - 2 A_{12}{}^2),$$
$$K_{12}{}^2 \colon \ 0 = (x^1)^{-2}(-A_{12}{}^1 + A_{22}{}^2).$$

We impose the resulting relations; set $A_{12}{}^2 = t$ and $A_{11}{}^2 = s$ to obtain

$$\tilde{\rho} = t^2 dx^1 \otimes dx^1, \quad X = \log(x^1) \partial_{x^2} \in \mathfrak{K}(\mathcal{N}), \quad \mathcal{N}(t, s) = \mathcal{N}(2t - 1, s, 0, t, 0, 0).$$

To ensure $\mathcal{N}(t, s)$ is not flat, we assume $t \neq 0$. We can apply Lemma 14.2 to assume $s = 0$ and obtain $\mathcal{N}_3^4(c_1)$ for $c_1 = t \neq 0$.

Case 2.3. Suppose $\alpha = 1$. We do not need to worry about x^2 divided by a power of x^1 since we have considered the case $\alpha < 0$ in Case 2.1. We first suppose there are no log terms so $X = (a_1 x^1 + a_2 x^2) \partial_{x^1} + (b_1 x^1 + b_2 x^2) \partial_{x^2}$. By subtracting an appropriate multiple of $x^1 \partial_{x^1} + x^1 \partial_{x^2}$, we may assume $b_2 = 0$. If $a_2 \neq 0$, we would obtain $\partial_{x^1} \in E(0)$, and we have already dealt with this possibility in Case 2.2. Thus we have $X = x^1(a_1 \partial_{x^1} + b_1 \partial_{x^2})$.

Case 2.3.1. Suppose $a_1 \neq 0$. We may assume $a_1 = 1$. We set $\tilde{x}^2 = x^2 - b_1 x^1$. We then have $\partial_{\tilde{x}^1} = \partial_{x^1} + b_1 \partial_{x^2}$ and $\partial_{\tilde{x}^2} = \partial_{x^2}$. We may therefore assume $b_1 = 0$. Consequently, we have that $K_{12}{}^1 \colon 0 = -(x^1)^{-1} A_{12}{}^1$ and $K_{22}{}^2 \colon 0 = -(x^1)^{-1} A_{22}{}^2$. Set $A_{12}{}^1 = 0$ and $A_{22}{}^2 = 0$ and obtain the relation $K_{11}{}^2 \colon 0 = (x^1)^{-1} A_{11}{}^2$. We have solved all the affine Killing equations so $x^1 \partial_{x^1} \in \mathfrak{K}(\mathcal{N})$. Set $A_{11}{}^1 = r$ and $A_{12}{}^2 = s$ to obtain

$$\tilde{\rho} = (1 + r - s)s \, dx^1 \otimes dx^1 \text{ and } \mathcal{N} = \mathcal{N}(r, 0, 0, s, 0, 0).$$

Depending on the values of $\{r, s\}$, this is either $\mathcal{N}_2^4(\kappa, \theta)$ for $\kappa = s$ and $\theta = r - (2s - 1) \neq 0$ or $\mathcal{N}_3^4(c_1)$ for $s = c_1$ and $r = 2s - 1$.

Case 2.3.2. Suppose $a_1 = 0$ so $X = x^1 \partial_{x^2}$. We obtain

$$K_{11}{}^1 : 0 = (x^1)^{-1} 2 A_{12}{}^1, \qquad\qquad K_{11}{}^2 : 0 = (x^1)^{-1}(-A_{11}{}^1 + 2A_{12}{}^2),$$
$$K_{12}{}^2 : 0 = (x^1)^{-1}(-A_{12}{}^1 + A_{22}{}^2).$$

We impose these relations and set $A_{12}{}^2 = c_2$ to obtain a solution with

$$\tilde{\rho} = (c_2 + c_2^2) dx^1 \otimes dx^1, \quad X = x^1 \partial_{x^2}, \quad \mathcal{N} = \mathcal{N}(2c_2, A_{11}{}^2, 0, c_2, 0, 0).$$

To ensure $\tilde{\rho} \neq 0$, we require $c_2 \notin \{0, -1\}$. If $A_{11}{}^2 \neq 0$, we can rescale x^2 to set $A_{11}{}^2 = 1$ and obtain $\mathcal{N}_1^4(c_2)$. If $A_{11}{}^2 = 0$, we obtain $\mathcal{N}_2^4(\kappa, \theta)$ for $\theta = 1$ and $\kappa = c_2$.

Case 2.3.3. Suppose we have $\log(x^1)$ terms. There cannot be negative powers of x^1. Any terms $x^2 \log(x^1)$ give rise to a $\log(x^1)$ in $E(0)$ which is false. Any terms with $x^1 \log(x^1)$ give rise to an x^1 term. This does not occur.

Case 2.4. Suppose $\alpha = 2$. Since there are no terms with $\Re(\alpha) < 0$, we need not worry about x^2 divided by a power of x^1. Nor need we worry about $\log(x^1)$ terms. Thus X is a pure quadratic polynomial. The analysis of Case 1.5 permits us to assume

$$X = (a_{11}(x^1)^2 + 2a_{12}x^1 x^2)\partial_{x^1} + (b_{11}(x^1)^2 + a_{12}(x^2)^2)\partial_{x^2}.$$

Case 2.4.1. Suppose $a_{12} = 0$ but $a_{11} = 1$. We have

$$K_{12}{}^1 : 0 = -A_{12}{}^1, \quad K_{12}{}^2 : 0 = A_{12}{}^2 - 2A_{12}{}^1 b_{11} + 2A_{22}{}^2 b_{11}, \quad K_{22}{}^2 : 0 = -A_{22}{}^2.$$

If $A_{12}{}^1 = 0$, $A_{22}{}^2 = 0$, and $A_{22}{}^1 = 0$, then $\rho = 0$. This case does not appear.

Case 2.4.2. Suppose $a_{12} = 0$, $a_{11} = 0$, and $b_{11} = 1$. We have

$$K_{11}{}^1 : 0 = 4A_{12}{}^1, \quad K_{11}{}^2 : 0 = 2 - 2A_{11}{}^1 + 4A_{12}{}^2, \quad K_{12}{}^2 : 0 = -2A_{12}{}^1 + 2A_{22}{}^2.$$

We solve these equations and set $A_{12}{}^2 = s$ to obtain an affine Killing vector field $(x^1)^2 \partial_{x^2}$ where $\tilde{\rho} = s(2 + s)dx^1 \otimes dx^1$ and $\mathcal{N} = \mathcal{N}(1 + 2s, A_{11}{}^2, 0, s, 0, 0)$. We can use Lemma 14.2 and take a shear with $b = A_{12}{}^2$ and $a = 1$ to assume $A_{11}{}^2 = 0$ and obtain $\mathcal{N}_2^4(s, 2)$; this was also obtained previously in Case 2.1.2.

Case 2.4.3. Suppose $a_{12} \neq 0$. We can assume $a_{12} = 1$ and perform a gauge transformation to assume $a_{11} = 0$ and obtain $X = 2x^1 x^2 \partial_{x^1} + (b_{11}(x^1)^2 + (x^2)^2)\partial_{x^2}$,

$$K_{22}{}^1 : 0 = 4A_{12}{}^1 - 2A_{22}{}^2 \quad \text{and} \quad K_{22}{}^2 : 0 = 2 + 4A_{12}{}^2.$$

We set $A_{12}{}^2 = -\frac{1}{2}$ and $A_{22}{}^2 = 2A_{12}{}^1$. We obtain $K_{12}{}^1 : 0 = 3 + 2A_{11}{}^1$ so $A_{11}{}^1 = -\frac{3}{2}$. We obtain $K_{11}{}^1 : 0 = -2A_{11}{}^2 + 4A_{12}{}^1 b_{11}$ and $K_{11}{}^2 : 0 = 3b_{11}$. Thus, $b_{11} = 0$ and $A_{11}{}^2 = 0$. We obtain a solution where $X = 2x^1 x^2 \partial_{x^1} + (x^2)^2 \partial_{x^2}$, $\mathcal{N} = \mathcal{N}(-\frac{3}{2}, 0, A_{12}{}^1, -\frac{1}{2}, 0, 2A_{12}{}^1)$ and

$$\tilde{\rho} = \begin{pmatrix} 0 & \frac{3}{2}A_{12}{}^1 \\ -\frac{3}{2}A_{12}{}^1 & (A_{12}{}^1)^2 \end{pmatrix}.$$

To ensure the geometry is not flat, we have $A_{12}{}^1 \neq 0$. We can therefore rescale x^2 to set $A_{12}{}^1 = 1$ and obtain $\mathcal{N}_2^3(0)$; this was missing in Case 1.5.1.

Case 2.5. We suppose $\alpha \neq 0, 1, 2$. Since $E(\beta) = \{0\}$ for $\Re(\beta < 0)$, there are no terms with a power of x^2 divided by a power of x^1. Similarly, we can assume there are no $\log(x^1)$ terms. Thus the analysis of Case 2.1 pertains. $\qquad\square$

14.2.4 RELATING TYPE $\mathcal{A}$ AND TYPE $\mathcal{B}$ MODELS.

The following result describes the intersection between the Type $\mathcal{A}$ and the Type $\mathcal{B}$ models; it is due to Brozos-Vázquez, García-Río, and Gilkey [7].

Theorem 14.17

1. *Let $\mathcal{N}$ be a non-flat Type $\mathcal{B}$ model. The following assertions are equivalent.*

   (a) $\mathcal{N}$ is locally affine isomorphic to a Type $\mathcal{A}$ model.

   (b) $\dim(\mathfrak{K}(\mathcal{N})) = 4$.

   (c) $\mathcal{N}$ is linearly isomorphic to $\mathcal{N}_1^4(c_2)$, $\mathcal{N}_2^4(\kappa, \theta)$, or $\mathcal{N}_3^4(c_1)$.

   (d) $A_{12}{}^1 = A_{22}{}^1 = A_{22}{}^2 = 0$.

2. *Let $\mathcal{M}$ be a non-flat Type $\mathcal{A}$ model. The following assertions are equivalent.*

   (a) $\mathcal{M}$ is locally affine isomorphic to a Type $\mathcal{B}$ model.

   (b) $\mathcal{M}$ is locally affine isomorphic to $\mathcal{M}_3^4(c_1)$ or to $\mathcal{M}_4^4(0)$.

Proof. We establish Assertion 1 as follows. Let $\mathcal{N}$ be a non-flat Type $\mathcal{B}$ model. If $\mathcal{N}$ is also a Type $\mathcal{A}$ model, then $\mathfrak{K}(\mathcal{N})$ contains both a 2-dimensional Abelian Lie subalgebra and a 2-dimensional non-Abelian Lie subalgebra. Consequently, $\dim(\mathfrak{K}(\mathcal{N})) \geq 3$. Since there are no Type $\mathcal{A}$ models with $\dim(\mathfrak{K}(\mathcal{N})) = 3$ and since $\mathcal{N}$ is non-flat, $\dim(\mathfrak{K}(\mathcal{N})) = 4$. Next suppose $\dim(\mathfrak{K}(\mathcal{N})) = 4$. By Theorem 14.16, $\mathcal{N}$ is linearly isomorphic to $\mathcal{N}_i^4(\cdot)$. Suppose $\mathcal{N}$ is linearly isomorphic to $\mathcal{N}_i^4(\cdot)$. By Lemma 14.11, $\mathcal{N}$ is also a Type $\mathcal{A}$ model. Thus Assertions 1-a, 1-b, and 1-c are equivalent. Suppose $\mathcal{N}$ is linearly isomorphic to $\mathcal{N}_i^4(\cdot)$. These models satisfy $A_{12}{}^1 = A_{22}{}^1 = A_{22}{}^2 = 0$. By Lemma 14.2, this condition is preserved by a shear. Consequently, Assertion 1-c implies Assertion 1-d. Conversely, suppose Assertion 1-d holds. We compute $\tilde{\rho} = A_{12}{}^2(1 + A_{11}{}^1 - A_{12}{}^2)dx^1 \otimes dx^1$ so $A_{12}{}^2 \neq 0$ and $1 + A_{11}{}^1 - A_{12}{}^2 \neq 0$.

Case 1. If $2A_{12}{}^2 - A_{11}{}^1 \neq 0$, perform a shear to set $A_{11}{}^2 = 0$. Set $A_{11}{}^1 = 2\kappa + \theta - 1$ and $A_{12}{}^2 = \kappa$. We have $\kappa \neq 0$ and $1 + A_{11}{}^1 - A_{12}{}^2 = \theta + \kappa$. If $\theta = 0$, we obtain $\mathcal{N}_3^4(c_1)$ for $c_1 = \kappa$. If $\theta \neq 0$, we obtain $\mathcal{N}_2^4(\kappa, \theta)$.

Case 2. If $2A_{12}{}^2 - A_{11}{}^1 = 0$, then we can rescale x^2 to assume either $A_{11}{}^2 = 1$ and obtain $\mathcal{N}_1^4(c_2)$ or that $A_{11}{}^2 = 0$ and obtain $\mathcal{N}_2^4(\kappa, 1)$.

This shows that Assertion 1-d implies Assertion 1-c. Assertion 2 is immediate from Assertion 1. $\qquad\square$

14.2.5 AFFINE KILLING COMPLETE.

Theorem 14.18   *Let $\mathcal{N}$ be a Type $\mathcal{B}$ model.*

1. *If $\dim(\mathfrak{K}(\mathcal{N})) = 2$ or if $\dim(\mathfrak{K}(\mathcal{N})) = 4$, then $\mathcal{N}$ is affine Killing complete.*

2. *If $\dim(\mathfrak{K}(\mathcal{N})) = 3$, then $\mathcal{N}$ is affine Killing complete if and only if $\mathcal{N}$ is linearly equivalent to the hyperbolic plane.*

3. *If $\dim(\mathfrak{K}(\mathcal{N})) = 6$, then $\mathcal{N}$ is affine Killing complete if and only if $\mathcal{N}$ is linearly equivalent to $\mathcal{N}_0^6$ or $\mathcal{N}_5^6$.*

Proof. If $\dim(\mathfrak{K}(\mathcal{N})) = 4$, then $\mathcal{N}$ is linearly equivalent to $\mathcal{N}_i^4(\cdot)$ for some i by Theorem 14.16. By Lemma 14.11, these manifolds are affine isomorphic to $\mathcal{M}_3^4(c_1)$, for some c_1, or to $\mathcal{M}_4^4(0)$; these manifolds are affine Killing complete by Lemma 13.16. If $\dim(\mathfrak{K}(\mathcal{N})) = 2$, then the vector fields in question are the Lie algebra of the $ax + b$ group and are complete. The Lie algebra of the hyperbolic plane $\mathcal{N}_4^3$ is the 3-dimensional group of isometries of $\mathcal{N}_4^3$; this is affine Killing complete. We will discuss this structure further in Section 14.3.

We complete the proof by examining the geometries $\mathcal{N}_i^3(\cdot)$ for $i = 1, 2, 3$.

Case 1. Suppose $\mathcal{N} = \mathcal{N}_1^3(\pm)$ or $\mathcal{N} = \mathcal{N}_2^3(c)$. Then $X = 2x^1x^2\partial_{x^1} + (x^2)^2\partial_{x^2}$ is an affine Killing vector field. We let $\sigma(t) = \left(\frac{1}{t^2}, -\frac{1}{t}\right)$. This is a flow curve for X, so this geometry is affine Killing incomplete.

Case 2. Suppose $\mathcal{N} = \mathcal{N}_3^3$. Then $X = 2x^1x^2\partial_{x^1} + ((x^1)^2 + (x^2)^2)\partial_{x^2}$ is an affine Killing vector field. We let $\sigma(t) = \left(\frac{1}{t^2-1}, -\frac{t}{t^2-1}\right)$. This is a flow curve for X, so this geometry is affine Killing incomplete. □

The structure $\mathcal{N}_3^3$ embeds in the pseudo-sphere, as we shall see in Section 14.3. Thus this geometry can be completed. We do not know if the remaining geometries $\mathcal{N}_1^3(\pm)$ and $\mathcal{N}_2^3(c)$ can be affine Killing completed.

14.2.6 INVARIANT AND PARALLEL TENSORS OF TYPE (1,1).

If $\mathcal{N}$ is of Type $\mathcal{B}$, then ∂_{x^2} and $x^1\partial_{x^1} + x^2\partial_{x^2}$ are affine Killing vector fields. We have

$$\mathcal{L}_{\partial_{x^2}}(\partial_{x^i} \otimes \partial_{x^j}) = 0 \quad \text{and} \quad \mathcal{L}_{x^1\partial_{x^1}+x^2\partial_{x^2}}(\partial_{x^i} \otimes \partial_{x^j}) = 0\,.$$

Thus T is invariant implies $T = T^i{}_j\partial_{x^i} \otimes dx^j$ where the coefficients $T^i{}_j$ are constant. We suppose $\text{Tr}\{T\} = 0$. We apply Theorem 14.16 and Theorem 14.17. If $\dim(\mathfrak{K}(\mathcal{N})) = 4$, then $\mathcal{N}$ is also Type $\mathcal{A}$ and the analysis in Section 13.2.8 pertains. However, it is worth doing an independent calculation. We have:

Example 14.19

1. $\mathfrak{K}(\mathcal{N}) = \text{span}\{x^1\partial_{x^1} + x^2\partial_{x^2}, \partial_{x^2}, x^1\partial_{x^1} - x^1\log(x^1)\partial_{x^2}, x^1\partial_{x^2}\}$. We have:

$$\mathcal{L}_{x^1\partial_{x^1}-x^1\log(x^1)\partial_{x^2}}T = \begin{pmatrix} -T^1{}_2(1+\log(x^1)) & -T^1{}_2 \\ 2T^1{}_1 + T^2{}_1 + 2T^1{}_1\log(x^1) & T^1{}_2(1+\log(x^1)) \end{pmatrix},$$

$$\mathcal{L}_{x^1\partial_{x^2}}T = \begin{pmatrix} T^1{}_2 & 0 \\ -2T^1{}_1 & -T^1{}_2 \end{pmatrix}.$$

2. $\mathfrak{K}(\mathcal{N}) = \text{span}\{x^1\partial_{x^1} + x^2\partial_{x^2}, \partial_{x^2}, x^1\partial_{x^1}, (x^1)^\alpha\partial_{x^2}\}$ for $\alpha \neq 0$. We have:

$$\mathcal{L}_{x^1\partial_{x^1}}T = \begin{pmatrix} 0 & -T^1{}_2 \\ T^2{}_1 & 0 \end{pmatrix} \quad \text{and} \quad \mathcal{L}_{(x^1)^\alpha\partial_{x^2}}T = \alpha(x^1)^{\alpha-1}\begin{pmatrix} T^1{}_2 & 0 \\ -2T^1{}_1 & -T^1{}_2 \end{pmatrix}.$$

3. $\mathfrak{K}(\mathcal{N}) = \text{span}\{x^1\partial_{x^1} + x^2\partial_{x^2}, \partial_{x^2}, x^1\partial_{x^1}, \log(x^1)\partial_{x^2}\}$. We have:

$$\mathcal{L}_{x^1\partial_{x^1}}T = \begin{pmatrix} 0 & -T^1{}_2 \\ T^2{}_1 & 0 \end{pmatrix} \quad \text{and}$$

$$\mathcal{L}_{\log(x^1)\partial_{x^2}}T = (x^1)^{-1}\begin{pmatrix} T^1{}_2 & 0 \\ -2T^1{}_1 & -T^1{}_2 \end{pmatrix}.$$

Consequently, there is no non-trivial invariant trace-free tensor of Type $(1,1)$ if $\dim(\mathfrak{K}(\mathcal{N})) = 4$. We now consider the situation when $\dim(\mathfrak{K}(\mathcal{N})) = 3$ and obtain a solution.

Example 14.20 $\mathfrak{K}(\mathcal{N}) = \text{span}\{x^1\partial_{x^1} + x^2\partial_{x^2}, \partial_{x^2}, 2x^1x^2\partial_{x^1} + ((x^2)^2 + \sigma(x^1)^2)\partial_{x^2}\}$ where we have $\sigma \in \{-1, 0, 1\}$. We compute:

$$\mathcal{L}_{2x^1x^2\partial_{x^1}+((x^2)^2+\sigma(x^1)^2)\partial_{x^2}}T = 2x^1\begin{pmatrix} T^1{}_2\sigma - T^2{}_1 & 2T^1{}_1 \\ -2T^1{}_1\sigma & T^2{}_1 - T^1{}_2\sigma \end{pmatrix}.$$

We set $T^1{}_1 = 0$ and $T^2{}_1 = T^1{}_2\sigma$. We have $\det(T) = -(T^1{}_2)^2\sigma$. There are three cases.

1. Suppose $\sigma = -1$. Taking $A_{11}{}^1 = -1$, $A_{11}{}^2 = 0$, $A_{12}{}^1 = 0$, $A_{12}{}^2 = -1$, $A_{22}{}^1 = 1$, and $A_{22}{}^2 = 0$ we have the Christoffel symbols of the metric of constant Gaussian curvature -1 on the hyperbolic plane. Take $T^1{}_2 = -1$ and $T^2{}_1 = 1$ to define the parallel complex structure on this Kähler surface.

2. Suppose $\sigma = 1$. For $A_{11}{}^1 = -1$, $A_{11}{}^2 = 0$, $A_{12}{}^1 = 0$, $A_{12}{}^2 = -1$, $A_{22}{}^1 = -1$, and $A_{22}{}^2 = 0$ we have the Christoffel symbols of the metric of constant Gaussian curvature on the Lorentzian hyperbolic plane. Take $T^1{}_2 = 1$ and $T^2{}_1 = 1$ to define the parallel para-complex structure; the hyperbolic plane is a para-Kähler surface.

3. Suppose $\sigma = 0$ so $T = \varepsilon\partial_{x^1} \otimes dx^2$. This gives an invariant nilpotent structure on the surfaces $\mathcal{N}_1^3(\pm)$ and $\mathcal{N}_2^3(c)$.

The following characterization of trace-free parallel $(1, 1)$-tensor fields on Type $\mathcal{B}$ models was obtained by Calviño-Louzao et al. [13]. It follows from Lemma 12.53.

Lemma 14.21   *If ∇ is a Type $\mathcal{B}$ connection on $M = \mathbb{R}^+ \times \mathbb{R}$ and if $\mathcal{P}^0(\mathcal{M}) \neq \{0\}$, then there exists $\alpha \in \mathbb{C}$ and $0 \neq \mathfrak{t} \in M_2^0(\mathbb{C})$ so that $(x^1)^\alpha \mathfrak{t} \in \mathcal{P}_{\mathbb{C}}^0(\mathcal{M})$.*

Lemma 13.3 shows that ρ_s is recurrent if and only if it is of rank 1 in the Type $\mathcal{A}$ setting. This fact is no longer true for Type $\mathcal{B}$ geometries.

Example 14.22   Let $\mathcal{M}$ be the Type $\mathcal{B}$ surface defined by setting $A_{22}{}^2 = (3 + 2\sqrt{3})/3$ and $A_{ij}{}^k = 1$ otherwise. We compute

$$\rho_s = \frac{1}{(x^1)^2} \begin{pmatrix} 1 + \frac{2}{\sqrt{3}} & \frac{1}{\sqrt{3}} \\ \frac{1}{\sqrt{3}} & \frac{2}{\sqrt{3}} - 1 \end{pmatrix}$$

and, consequently, ρ_s has rank 1. Assume $\dim(\mathcal{P}^0(\mathcal{M})) \geq 1$. It follows from Lemma 14.21 that there exists an element in $\mathcal{P}_{\mathbb{C}}^0(\mathcal{M})$ of the form $T = (x^1)^\alpha (\mathfrak{t}^i{}_j)$ where $0 \neq (\mathfrak{t}^i{}_j) \in M_2^0(\mathbb{C})$. Setting $T^i{}_{j;2} = 0$ yields the relations:

$$(x^1)^{\alpha-1} \begin{pmatrix} \mathfrak{t}^2{}_1 - \mathfrak{t}^1{}_2 & -2\mathfrak{t}^1{}_1 - \frac{2}{\sqrt{3}}\mathfrak{t}^1{}_2 \\ 2\mathfrak{t}^1{}_1 + \frac{2}{\sqrt{3}}\mathfrak{t}^2{}_1 & \mathfrak{t}^1{}_2 - \mathfrak{t}^2{}_1 \end{pmatrix} = \begin{pmatrix} 0 & 0 \\ 0 & 0 \end{pmatrix}.$$

We solve this relation to see $\mathfrak{t}^2{}_1 = \mathfrak{t}^1{}_2$ and $\mathfrak{t}^1{}_1 = -\frac{1}{\sqrt{3}}\mathfrak{t}^1{}_2$. Substituting these relations and setting $T^i{}_{j;1} = 0$ then yields:

$$(x^1)^{\alpha-1} \begin{pmatrix} -\frac{\alpha}{\sqrt{3}}\mathfrak{t}^1{}_2 & \left(\alpha + \frac{2}{\sqrt{3}}\right)\mathfrak{t}^1{}_2 \\ \left(\alpha - \frac{2}{\sqrt{3}}\right)\mathfrak{t}^1{}_2 & \frac{\alpha}{\sqrt{3}}\mathfrak{t}^1{}_2 \end{pmatrix} = \begin{pmatrix} 0 & 0 \\ 0 & 0 \end{pmatrix}.$$

This shows $\mathfrak{t}^1{}_2 = 0$, so $T = 0$. This shows $\mathcal{P}^0(\mathcal{M})$ is trivial.

14.3   SYMMETRIC SPACES

In Section 14.3.1 (see Lemma 14.23), we will show that up to linear equivalence, the only non-flat Type $\mathcal{B}$ models which are affine symmetric spaces are given by

1. $\mathcal{N}_1^4(c_2 = -\frac{1}{2})$.
2. $\mathcal{N}_2^4(\kappa = c, \theta = -2c)$ for $c \neq 0$.
3. $\mathbb{L} := \mathcal{N}_3^3$ (the Lorentzian hyperbolic plane).
4. $\mathbb{H} := \mathcal{N}_4^3$ (the hyperbolic plane).

We have $\mathcal{N}_2^4(c, -2c)$ for $c \neq 0$, and $\mathcal{N}_3^4(-\frac{1}{2})$ are affine diffeomorphic to the affine symmetric space $\mathcal{M}_3^4(-\frac{1}{2})$; this Type $\mathcal{A}$ geometry was studied in Chapter 13 and up to linear equivalence, the only other Type $\mathcal{A}$ geometry which is an affine symmetric space and non-flat is $\mathcal{M}_5^4(0)$, which is not a Type $\mathcal{B}$ geometry. In Section 14.3.2 (see Theorem 14.24), we will show that the only Type $\mathcal{B}$ geometries with $\dim(\mathfrak{K}(\mathcal{N})) > 2$ which are geodesically complete are affine symmetric spaces; $\mathbb{L}$ is not geodesically complete while $\mathcal{N}_2^4(c, -2c), \mathcal{N}_3^4(-\frac{1}{2})$, and $\mathbb{H}$ are geodesically complete.

In Section 14.3.3, we discuss the geometry of spaces of constant sectional curvature and generalize results concerning $\mathfrak{H}$ and $\mathfrak{L}$ to this setting, in Section 14.3.4, we will study the geometry of the hyperbolic plane $\mathbb{H}$, and in Section 14.3.5, we study the hyperbolic pseudo-sphere $\mathfrak{H}$, which is a different but equivalent model for hyperbolic geometry. In Section 14.3.6, we treat the geometry of the Lorentzian hyperbolic plane $\mathbb{L}$, and in Section 14.3.7, we present the pseudo-sphere $\mathfrak{L}$ and its universal cover $\tilde{\mathfrak{L}}$; $\mathfrak{L}$ and $\tilde{\mathfrak{L}}$ are geodesically complete. The Lorentzian hyperbolic plane is geodesically incomplete but admits a proper affine embedding in $\mathfrak{L}$.

14.3.1 CLASSIFICATION OF TYPE $\mathcal{B}$ SYMMETRIC SPACES.
It follows from Lemma 13.8 and Theorem 13.22 that any Type $\mathcal{A}$ model which is a non-flat affine symmetric space is linearly equivalent to $\mathcal{M}_2^4(-\frac{1}{2})$, $\mathcal{M}_3^4(-\frac{1}{2})$, or $\mathcal{M}_5^4(0)$. There is a similar classification here.

Lemma 14.23   *If $\mathcal{N}$ is a non-flat Type $\mathcal{B}$ model, then $\mathcal{N}$ is an affine symmetric space if and only if $\mathcal{N}$ is linearly equivalent to $\mathcal{N}_1^4(-\frac{1}{2})$, $\mathcal{N}_2^4(c, -2c)$ for $c \neq 0$, $\mathbb{L} = \mathcal{N}_3^3$, or $\mathbb{H} = \mathcal{N}_4^3$.*

Proof. Let $\mathcal{N}$ be a Type $\mathcal{B}$ model so $\nabla\rho = 0$. By Theorem 12.9, ρ is symmetric. Consequently, $A_{12}{}^1 + A_{22}{}^2 = 0$. Let $\tilde{\rho} := (x^1)^2\rho$ and $\bar{\rho}_{ij;k} := (x^1)^3\rho_{ij;k}$. We examine the relations which arise from setting $\bar{\rho}_{ij;k} = 0$.

Case 1. Suppose $A_{22}{}^1 \neq 0$. We rescale to set $A_{22}{}^1 = \varepsilon$ for $\varepsilon = \pm 1$ and then, by Lemma 14.2, perform a shear to ensure $A_{12}{}^1 = 0$. Since $A_{12}{}^1 + A_{22}{}^2 = 0$, we have $A_{22}{}^2 = 0$ as well. We have that $\bar{\rho}_{22;2}: 0 = 2A_{11}{}^2$ and $\bar{\rho}_{21;2}: 0 = 2A_{12}{}^2(A_{12}{}^2 - A_{11}{}^1)\varepsilon$. We set $A_{11}{}^2 = 0$ and consider subcases.

Case 1.1. Suppose $A_{12}{}^2 \neq 0$. Then $A_{11}{}^1 = A_{12}{}^2$. We obtain $\bar{\rho}_{22;1}: 0 = 2(1 + A_{12}{}^2)\varepsilon$. This yields $A_{12}{}^2 = -1$. We obtain $\mathcal{N} = \mathcal{N}(-1, 0, 0, -1, \varepsilon, 0)$, which is $\mathbb{L} = \mathcal{N}_3^3$ if $\varepsilon = -1$ and is $\mathbb{H} = \mathcal{N}_4^3$ if $\varepsilon = +1$.

Case 1.2. Suppose $A_{12}{}^2 = 0$. We obtain $\bar{\rho}_{22;1}: 0 = 2(1 - A_{11}{}^1)\varepsilon$. This yields $A_{11}{}^1 = 1$. We then obtain $\rho = 0$ contrary to our assumption. This case does not occur.

Case 2. Suppose $A_{22}{}^1 = 0$ and $A_{12}{}^1 \neq 0$. We then obtain $\bar{\rho}_{22;2}: 0 = -4(A_{12}{}^1)^3$, so this case is impossible.

Case 3. We assume $A_{22}{}^1 = 0$ and $A_{12}{}^1 = 0$ to obtain $A_{22}{}^2 = 0$ as well. The remaining non-zero component of $\nabla\rho$ is $\bar{\rho}_{11;1}: 0 = -2(1 + A_{11}{}^1)(1 + A_{11}{}^1 - A_{12}{}^2)A_{12}{}^2$. Since $\mathcal{N}$ is not flat,

$\tilde{\rho} = (1 + A_{11}{}^1 - A_{12}{}^2)A_{12}{}^2 dx^1 \otimes dx^1 \neq 0$. Thus we have that $A_{11}{}^1 = -1$ and, consequently, we have $\mathcal{N} = \mathcal{N}(-1, A_{11}{}^2, 0, A_{12}{}^2, 0, 0)$. If $A_{12}{}^2 = -\frac{1}{2}$ and $A_{11}{}^2 \neq 0$, we can rescale x^2 to set $A_{11}{}^2 = 1$ and obtain $\mathcal{N}(-1, 1, 0, -\frac{1}{2}, 0, 0) = \mathcal{N}_1^4(-\frac{1}{2})$. If $c = A_{12}{}^2 \neq -\frac{1}{2}$, apply a shear to set $A_{11}{}^2 = 0$ and obtain $\mathcal{N}(-1, 0, 0, c, 0, 0) = \mathcal{N}_2^4(\kappa = c, \theta = -2c)$; we assume $c \neq 0$ to ensure that $\rho \neq 0$. $\square$

A similar argument can be used to analyze the situation when the surface in question has torsion; we refer to D'Ascanio, Gilkey, and Pisani [23] as the analysis is beyond the scope of the current volume.

14.3.2 GEODESIC COMPLETENESS. In Section 13.3.5, we determined which Type $\mathcal{A}$ models were geodesically complete. The situation in the Type $\mathcal{B}$ setting is considerably more difficult and the answer in general is not known. We apply Theorem 14.16 to see that any Type $\mathcal{B}$ model with $\dim(\mathfrak{K}(\mathcal{N})) > 2$ is linearly isomorphic to one of the models $\mathcal{N}_i^{\nu}(\cdot)$ of Definition 14.1 for $\nu = 3, 4, 6$. We examine the situation for these examples as follows.

Theorem 14.24   *Let $\mathcal{N}$ be a Type $\mathcal{B}$ model with $\dim(\mathfrak{K}(\mathcal{N})) > 2$.*

1. *$\mathcal{N}_5^6$ is geodesically complete and $\mathcal{N}_i^6(\cdot)$ is geodesically incomplete for $i \neq 5$.*

2. *$\mathcal{N}_1^4(-\frac{1}{2})$ and $\mathcal{N}_2^4(c, -2c)$ for $c \neq 0$ are geodesically complete. $\mathcal{N}_i^4(\cdot)$ is essentially geodesically incomplete otherwise.*

3. *$\mathcal{N}_1^3(\pm)$ and $\mathcal{N}_2^3(c)$ are essentially geodesically incomplete.*

4. *$\mathbb{L}$ is geodesically incomplete and $\mathbb{H}$ is geodesically complete.*

Remark 14.25   As in the Type $\mathcal{A}$ setting, all of the examples of Type $\mathcal{B}$ models which are geodesically complete, non-flat, and with $\dim(\mathfrak{K}(\mathcal{N})) > 2$ are affine symmetric spaces. The geodesically incomplete Type $\mathcal{A}$ models which are symmetric spaces can be completed; similarly we shall see in Section 14.3.7 that the Lorentzian hyperbolic plane $\mathbb{L}$ can be completed by embedding it in the pseudo-sphere $\mathfrak{L}$. As noted previously, $\mathcal{N}_1^4(-\frac{1}{2})$ and $\mathcal{N}_2^4(c, -2c)$ are affine equivalent to $\mathcal{M}_3^4(-\frac{1}{2})$; this geometry was studied in Chapter 13.

Proof. Suppose first $\mathcal{N}$ is flat. By Lemma 14.11, the maps Ψ_i^6 given in Definition 14.10 are affine embeddings of $\mathcal{N}_i^6(\cdot)$ in $\mathcal{M}_0^6$. They are not surjective for $i \neq 5$. Consequently, these geometries are geodesically incomplete. The map Ψ_5^6 is an affine diffeomorphism. Therefore, $\mathcal{N}_5^6$ is geodesically complete; Assertion 1 follows. Suppose next $\mathcal{N}$ is linearly isomorphic to $\mathcal{N}_i^4(\cdot)$. We have $\mathcal{N}_1^4(c_2)$ is affine isomorphic to $\mathcal{M}_3^4(c_2)$, $\mathcal{N}_2^4(\kappa, \theta)$ is linearly isomorphic to $\mathcal{M}_3^4(\frac{\kappa}{\theta})$, and $\mathcal{N}_3^4(c_1)$ is linearly isomorphic to $\mathcal{M}_4^4(0)$. By Lemma 13.30, $\mathcal{M}_3^4(c_1)$ is essentially geodesically incomplete for $c_1 \neq -\frac{1}{2}$, while $\mathcal{M}_3^4(-\frac{1}{2})$ is an affine symmetric space which is geodesically complete (see Lemma 13.8). Furthermore, $\mathcal{M}_4^4(0)$ is essentially geodesically incomplete. Assertion 2 follows. Finally, assume $\dim(\mathfrak{K}(\mathcal{N})) = 3$.

Case 1. Let $\mathcal{N} = \mathcal{N}_1^3(\pm) = \mathcal{N}(-\frac{3}{2}, 0, 0, -\frac{1}{2}, \mp\frac{1}{2}, 0)$ or $\mathcal{N} = \mathcal{N}_2^3(c) = \mathcal{N}(-\frac{3}{2}, 0, 1, -\frac{1}{2}, c, 2)$. Let $\sigma(t) = (t^{-2}, 0)$. Then $\dot{\sigma} = t^{-3}(-2, 0)$ and $\ddot{\sigma} = t^{-4}(6, 0)$. The only non-trivial geodesic equation is $\ddot{x}^1 - \frac{3}{2}(x^1)^{-1}\dot{x}^1\dot{x}^1 = 6t^{-4} - 4\frac{3}{2}t^2 t^{-3}t^{-3} = 0$. Thus these geometries are geodesically incomplete. Suppose there is a geodesically complete geometry $\tilde{\mathcal{N}}$ modeled on $\mathcal{N}_1^3(\pm)$ or on $\mathcal{N}_2^3(c)$. We may suppose without loss of generality that $\tilde{\mathcal{N}}$ is simply connected. Copy a small part of σ into $\tilde{\mathcal{N}}$. Extend σ to a globally defined affine geodesic τ. By Lemma 14.15, $\mathfrak{K}(\mathcal{N}) = \mathrm{span}\{x^1\partial_{x^1} + x^2\partial_{x^2}, \partial_{x^2}, 2x^1x^2\partial_{x^1} + (x^2)^2\partial_{x^2}\}$. By Theorem 12.52, $\tilde{\mathcal{N}}$ is real analytic. Choose global affine Killing vector fields on $\tilde{\mathcal{N}}$ so that near σ we have

$$\xi_1 = x^1\partial_{x^1} + x^2\partial_{x^2}, \quad \xi_2 = \partial_{x^2}, \quad \xi_3 = 2x^1x^2\partial_{x^1} + (x^2)^2\partial_{x^2}.$$

Since ξ_3 vanishes on σ and since the structures are real analytic, ξ_3 vanishes identically on τ. Consequently, $\{\xi_1, \xi_2\}$ forms a frame for the tangent bundle along $\tilde{\sigma}$. Thus we can express $\dot{\tau} = \kappa^1(t)\xi_1(t) + \kappa^2(t)\xi_2(t)$ where the κ^i are now real analytic functions defined for $t \in \mathbb{R}$. Since $\kappa^1(t) = -2t^{-1/2}$, this is impossible. Thus the structures $\mathcal{N}_1^3(\pm)$ and $\mathcal{N}_2^3(c)$ are essentially geodesically incomplete.

Case 2. We show $\mathbb{L} = \mathcal{N}(-1, 0, 0, -1, -1, 0)$ is geodesically incomplete by verifying that the geodesic equations are satisfied by the curve $\sigma(t) = t^{-1}(1, 1)$: $x^1\ddot{x}^1 - \dot{x}^1\dot{x}^1 - \dot{x}^2\dot{x}^2 = 0$ and $x^1\ddot{x}^2 - 2\dot{x}^1\dot{x}^2 = 0$. We will show that $\mathbb{H}$ is geodesically complete in Section 14.3.4.   □

We give below the geodesic structure of these geometries with basepoint $(1, 0)$. The geodesics of the hyperbolic plane $\mathcal{N}_4^3$ are circles perpendicular to the vertical axis that go through $(1, 0)$; the geodesic structure of the remaining models are more complicated. Note that none of the geodesics ever reach the asymptote $x^1 = 0$.

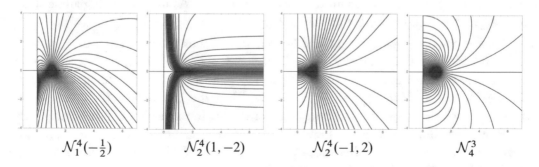

$\mathcal{N}_1^4(-\frac{1}{2})$ $\mathcal{N}_2^4(1, -2)$ $\mathcal{N}_2^4(-1, 2)$ $\mathcal{N}_4^3$

14.3.3 THE GEOMETRY OF PSEUDO-SPHERES.

Before discussing the hyperbolic plane $\mathbb{H}$ and the Lorentzian hyperbolic plane $\mathbb{L}$, it is convenient to digress and discuss *pseudo-spheres* in complete generality. These provide models for spaces of constant sectional curvature. We adopt the following notational conventions. Let $\varepsilon := (\varepsilon_{ij})$ where $\varepsilon_{ij} = 0$ if $i \neq j$ and $\varepsilon_{ij} = \pm 1$ if $i = j$. We shall assume that $\varepsilon_{m+1, m+1} = +1$ to simplify certain sign conventions.

For example, we will take

$$\varepsilon_{\mathcal{S}} := \begin{pmatrix} 1 & 0 & 0 \\ 0 & 1 & 0 \\ 0 & 0 & 1 \end{pmatrix}, \; \varepsilon_{\mathfrak{L}} := \begin{pmatrix} -1 & 0 & 0 \\ 0 & 1 & 0 \\ 0 & 0 & 1 \end{pmatrix}, \; \varepsilon_{\mathfrak{H}} := \begin{pmatrix} -1 & 0 & 0 \\ 0 & -1 & 0 \\ 0 & 0 & 1 \end{pmatrix} \quad (14.3.\text{a})$$

to discuss the round sphere, a model $\mathfrak{L}$ for the Lorentzian hyperbolic plane $\mathbb{L}$, and a model $\mathfrak{H}$ for the hyperbolic plane $\mathbb{H}$; we obtain a metric on $\mathfrak{H}$ of signature $(2,0)$ rather than the usual Riemannian metric of signature $(0, 2)$, but this does not change the underlying affine structure.

Let signature$(\varepsilon) = (p, q)$ where p is the number of $+1$ and q the number of -1 entries. Set $\langle \cdot, \cdot \rangle_\varepsilon := \varepsilon_{ij} dx^i \otimes dx^j$. Let the *pseudo-sphere*

$$\mathcal{S}_\varepsilon := \{ \vec{x} \in \mathbb{R}^{m+1} : \langle \vec{x}, \vec{x} \rangle_\varepsilon = +1 \} \quad (14.3.\text{b})$$

given the induced pseudo-Riemannian metric g_ε of signature $(q, p - 1)$. Denote the associated *orthogonal group* by $\mathcal{O}_\varepsilon := \{ T \in \mathrm{GL}(m + 1, \mathbb{R}) : T^* \langle \cdot, \cdot \rangle_\varepsilon = \langle \cdot, \cdot \rangle_\varepsilon \}$. Let g_ε be the restriction of $\langle \cdot, \cdot \rangle_\varepsilon$ to the tangent bundle of $\mathcal{S}_\varepsilon$. We summarize as follows the basic facts concerning pseudo-spheres.

Lemma 14.26 $(\mathcal{S}_\varepsilon, g_\varepsilon)$ *is a pseudo-Riemannian manifold of signature* $(p, q - 1)$ *on which* $\mathcal{O}_\varepsilon$ *acts transitively. If T is the germ of an isometry of $(\mathcal{S}_\varepsilon, g_\varepsilon)$, then $T \in \mathcal{O}_\varepsilon$. $(\mathcal{S}_\varepsilon, g_\varepsilon)$ is an affine symmetric space, $R(x, y, z, w) = \langle x, w \rangle_\varepsilon \langle y, z \rangle_\varepsilon - \langle x, z \rangle_\varepsilon \langle y, w \rangle_\varepsilon$, $\rho_\varepsilon = (m - 1) g_\varepsilon$, and $(\mathcal{S}_\varepsilon, g_\varepsilon)$ has constant sectional curvature $+1$. $(\mathcal{S}_\varepsilon, g_\varepsilon)$ is geodesically complete and any pseudo-Riemannian manifold of constant sectional curvature $+1$ is locally isometric to $(\mathcal{S}_\varepsilon, g_\varepsilon)$ for some ε.*

By replacing ε by $-\varepsilon$ and considering $\{ \vec{x} : \langle \vec{x}, \vec{x} \rangle = -1 \}$, we can obtain models for pseudo-Riemannian manifolds with constant sectional curvature -1. As the underlying affine structure is unchanged, we need not deal with this setting.

Proof. If $\vec{x} \in \mathcal{S}_\varepsilon$, then $\vec{x}$ is spacelike and $T_{\vec{x}}(\mathcal{S}_\varepsilon) = \vec{x}^\perp$. Thus the restriction of $\langle \cdot, \cdot \rangle_\varepsilon$ to $T_{\vec{x}} \mathcal{S}_\varepsilon$ defines a pseudo-Riemannian metric g_ε on $\mathcal{S}_\varepsilon$ of signature $(p, q - 1)$. The orthogonal group $\mathcal{O}_\varepsilon$ preserves the structures and acts isometrically on $(\mathcal{S}_\varepsilon, g_\varepsilon)$. Let $\xi_{m+1} \in \mathcal{S}_\varepsilon$. Choose an orthonormal basis $\{\xi_1, \dots, \xi_m\}$ for $\xi_{m+1}^\perp$ so $\langle \xi_i, \xi_j \rangle = \varepsilon_{ij}$. Then the matrix $T = (\xi_1, \dots, \xi_{m+1}) \in \mathcal{O}_\varepsilon$ and $T e_i = \xi_i$ where $\{e_i\}$ is the standard basis for $\mathbb{R}^{m+1}$. In particular, $\mathcal{O}_\varepsilon$ acts transitively on $\mathcal{S}_\varepsilon$ by isometries. Suppose T is an isometry of some component of $\mathcal{S}_\varepsilon$. Let $\{e_1, \dots, e_{m+1}\}$ be the standard basis for $\mathbb{R}^{m+1}$. Then $\xi_{m+1} := T e_{m+1}$ is a space-like unit vector. Furthermore, $\xi_1 := T_*(e_1)$, ..., $\xi_m := T_*(e_m)$ is a basis for $T_{\xi_{m+1}} \mathcal{S}_\varepsilon = \xi_{m+1}^\perp$ with $\langle \xi_i, \xi_j \rangle_\varepsilon = \langle e_i, e_j \rangle_\varepsilon$. Let $\tilde{T}$ be the matrix with columns $(\xi_1, \dots, \xi_{m+1})$. Then $\tilde{T} \in \mathcal{O}_\varepsilon$. We have $\tilde{T}^{-1} T e_{m+1} = e_{m+1}$ and $(\tilde{T}^{-1} T)_* e_i = e_i$ for $1 \le i \le m$. Thus $\tilde{T}^{-1} T$ preserves e_{m+1} and is the identity on $T_{e_{m+1}} \mathcal{S}_\varepsilon$. Since $\tilde{T}^{-1} T$ is the germ of an isometry of $\mathcal{S}_\varepsilon$ it follows $\tilde{T}^{-1} T$ is the identity. Therefore $T = \tilde{T} \in \mathcal{O}_\varepsilon$. This is, of course, only a local result. Global isometries may be represented by different elements of the orthogonal group on different arc components of $\mathcal{S}_\varepsilon$. As this will play no role in our analysis, we ignore this point.

Let $T(e_i) = -e_i$ for $1 \leq i \leq m$ and $Te_{m+1} = e_{m+1}$ define an element of $\mathcal{O}_\varepsilon$. Then $Te_{m+1} = e_{m+1}$ and $T_* = -\mathrm{Id}$ on $T_{e_{m+1}}\mathcal{S}_\varepsilon = e_{m+1}^\perp$. Since $(\mathcal{S}_\varepsilon, g_\varepsilon)$ is a homogeneous space, it follows that $(\mathcal{S}_\varepsilon, g_\varepsilon)$ is a global affine symmetric space. Since $(\mathcal{S}_\varepsilon, g_\varepsilon)$ is a homogeneous space, it suffices to compute the curvature tensor at a single point. We choose the point e_{m+1}. We have $\varepsilon_{11}(x^1)^2 + \cdots + \varepsilon_{mm}(x^m)^2 + (x^{m+1})^2 = 1$. Thus it is natural to introduce

$$\Phi(u^1, \ldots, u^m) = (u^1, \ldots, u^m, (1 - \varepsilon_{11}(u^1)^2 - \cdots - \varepsilon_{mm}(u^m)^2)^{1/2}),$$
$$\Phi_*(\partial_{u^i}) = e_i - \varepsilon_{ii}u^i(1 - \varepsilon_{11}(u^1)^2 - \cdots - \varepsilon_{mm}(u^m)^2)^{-1/2}e_{m+1}. \qquad (14.3.c)$$

We then have

$$g_{ij} = \varepsilon_{ij} + \varepsilon_{ii}\varepsilon_{jj}u^iu^j + O(\|u\|^3) \quad \text{for} \quad 1 \leq i, j \leq m. \qquad (14.3.d)$$

In the proof of Lemma 3.8 of Book I, we showed that if all the derivatives of the metric vanish at the origin, then

$$R_{ijk\ell}(0) = \tfrac{1}{2}\{g_{j\ell/ik} + g_{ik/j\ell} - g_{jk/i\ell} - g_{i\ell/jk}\}(0). \qquad (14.3.e)$$

Combining Equation (14.3.d) and Equation (14.3.e) then determines R; the calculation of the Ricci tensor is then immediate.

A curve σ in a non-degenerate hypersurface $\mathcal{S}$ in $(\mathbb{R}^m, \langle \cdot, \cdot \rangle_\varepsilon)$ is an affine geodesic if and only if $\ddot{\sigma} \perp \mathcal{S}$. We use this observation in what follows. Since $\mathcal{S}_\varepsilon$ is a homogeneous space, it suffices to consider an affine geodesic σ that starts from e_{m+1}. Let $\xi := \dot{\sigma}(0)$. There are three cases to be considered. Suppose first ξ is timelike. We may assume $\langle \xi, \xi \rangle_\varepsilon = -1$ and assume the coordinates on $\mathbb{R}^{m+1}$ are chosen so that $\xi = e_1$ where

$$\langle \cdot, \cdot \rangle = -dx^1 \otimes dx^1 \pm dx^2 \otimes dx^2 \pm \cdots \pm dx^m \otimes dx^m + dx^{m+1} \otimes dx^{m+1}.$$

Let $\sigma(t) := \sinh(t)e_1 + \cosh(t)e_{m+1}$. We verify that σ is an affine geodesic by checking that $\langle \sigma(t), \sigma(t) \rangle_\varepsilon = 1$, $\langle \sigma(t), \dot{\sigma}(t) \rangle_\varepsilon = 0$, and $\langle \dot{\sigma}(t), \dot{\sigma}(t) \rangle_\varepsilon = -1$. This implies σ is a timelike affine geodesic starting from e_{m+1} with initial direction e_1 which exists for all time. If ξ is spacelike, we assume the coordinates on $\mathbb{R}^{m+1}$ are chosen so $\xi = e_1$ where

$$\langle \cdot, \cdot \rangle = dx^1 \otimes dx^1 \pm dx^2 \otimes dx^2 \pm \cdots \pm dx^m \otimes dx^m + dx^{m+1} \otimes dx^{m+1}.$$

The remainder of the analysis is the same if we set $\sigma(t) = \sin(t)e_1 + \cos(t)e_{m+1}$. Finally, if ξ is null, assume $\xi = e_1 + e_2$ where

$$\langle \cdot, \cdot \rangle = dx^1 \otimes dx^1 - dx^2 \otimes dx^2 \pm \cdots \pm dx^m \otimes dx^m + dx^{m+1} \otimes dx^{m+1}.$$

Set $\sigma(t) = t(e_1 + e_2) + e_{m+1}$. We verify that σ is an affine geodesic by checking that $\langle \sigma(t), \sigma(t) \rangle_\varepsilon = 1$, $\langle \sigma(t), \dot{\sigma}(t) \rangle_\varepsilon = 0$, and $\langle \dot{\sigma}(t), \dot{\sigma}(t) \rangle_\varepsilon = 0$. Consequently, σ is a null affine geodesic which is defined for all time.

Any pseudo-Riemannian manifold of constant sectional curvature is locally symmetric. We apply Lemma 7.15 of Book II, which showed that a pseudo-Riemannian symmetric space is characterized up to local isometry by the form of its curvature tensor and metric at a single point. □

Let ∇_ε be the Levi-Civita connection of the pseudo-Riemannian manifold. Denote the underlying affine structure by $\mathcal{M}_\varepsilon := (\mathcal{S}_\varepsilon, \nabla_\varepsilon)$. Since the Levi-Civita connection of g_ε is the same as the Levi-Civita connection of $-g_\varepsilon$, the sign of the sectional curvature is irrelevant. Let $X_{ij} := \varepsilon_{ii} x^i \partial_{x^j} - \varepsilon_{jj} x^j \partial_{x^i}$ where we do not sum over repeated indices. If $f(x) = \langle x, x \rangle$, then $X_{ij}(f) = 0$. Thus the vector fields X_{ij} are tangential to $\mathcal{S}_\varepsilon$.

Lemma 14.27   Let $\mathcal{M}_\varepsilon$ be the underlying affine manifold defined by the pseudo-sphere $\mathcal{S}_\varepsilon$ of Equation (14.3.b). Then $\mathfrak{A}(\mathcal{M}_\varepsilon) = \mathrm{span}_{i<j}\{X_{ij}\}$, there are no affine gradient Ricci solitons on $\mathcal{M}_\varepsilon$, $\mathcal{M}_\varepsilon$ is strongly projectively flat, and

$$
E(\mu, \mathcal{M}_\varepsilon) = \left\{ \begin{array}{ll} \mathrm{span}\{x^1, \ldots, x^{m+1}\} & \text{if } \mu = -\frac{1}{m-1} \\ \mathrm{span}\{\mathbb{1}\} & \text{if } \mu = 0 \\ \{0\} & \text{otherwise} \end{array} \right\}.
$$

Proof. We apply Lemma 14.26. Let X be an affine Killing vector field and let Φ_t^X be the 1-parameter family of local diffeomorphisms corresponding to X. Then Φ_t^X commutes with ∇_ε, and hence preserves the Ricci tensor. We have $\rho_\varepsilon = (m-1)g_\varepsilon$. Consequently, the Φ_t^X are isometries, and thus X is an affine Killing vector field. The $\{X_{ij}\}$ represent the action of the Lie algebra of $\mathcal{O}_\varepsilon$ on $\mathbb{R}^{m+1}$ and, consequently, $\mathfrak{A}(\mathcal{M}_\varepsilon) = \mathrm{span}_{i<j}\{X_{ij}\}$.

Suppose ψ is an affine gradient Ricci soliton on $\mathcal{M}_\varepsilon$. Since $\nabla_\varepsilon \rho_\varepsilon = 0$, Lemma 12.12 shows $R_{ij} d\psi = 0$ for all i, j. We then have $R_{ij} e^k = \pm(\delta_{ik} e^j - \delta_{jk} e^i)$. Consequently, we have that $R_{ij} d\psi = 0$ for all i, j implies $d\psi = 0$ so $\psi = \mathbb{1}$. This implies $\rho_\varepsilon = 0$, which is false. Consequently, there are no affine gradient Ricci solitons on $\mathcal{M}_\varepsilon$.

Suppose $\mu = -\frac{1}{m-1}$. We wish to show $E(\mu, \mathcal{M}_\varepsilon) = \mathrm{span}\{x^1, \ldots, x^{m+1}\}$. Since $\mathcal{M}_\varepsilon$ is a homogeneous space, it suffices to show that, at the point e_{m+1}, the coordinate functions $\{x^1, \ldots, x^{m+1}\}$ solve the affine quasi-Einstein equation. We choose the coordinate system of Equation (14.3.c). At the origin, all the first derivatives of the metric and the Christoffel symbols vanish. We have $x^i = u^i$ for $1 \le i \le m$. Consequently, $\mathcal{H}(x^i)(0) = 0$ and $x^i \rho_\varepsilon(0) = 0$ and the affine quasi-Einstein equation is satisfied. Because

$$
x^{m+1} = (1 - \varepsilon_i (u^i)^2)^{\frac{1}{2}} \text{ and } \mathcal{H}(x^{m+1})(0) = \partial_{u^i} \partial_{u^j} (1 - \varepsilon_i (u^i)^2)^{\frac{1}{2}}(0),
$$

we have that $\mathcal{H}_{ij}(x^{m+1})(0) = -\varepsilon_{ij} = -\frac{1}{m-1}(-(m-1))\varepsilon_{ij} = -\frac{1}{m-1} x^{m+1} (\rho_\varepsilon)_{ij}(0)$. Suppose next that μ is not the critical eigenvalue, i.e., that $\mu(m-1) \ne -1$. If $\{\mathcal{H}f\}_{ij} = \mu f (\rho_\varepsilon)_{ij}$, then we use the fact that $(f\rho_\varepsilon)_{ij;k} = f_{;k}(\rho_\varepsilon)_{ij}$ and the identity $h_{;ijk} - h_{;ikj} = \{R_{kj}(dh)\}_i$ to equate

$$
\{R_{kj}(dh)\}_i = R_{jki}{}^\ell f_{;\ell} = \varepsilon_{ik} f_{;j} - \varepsilon_{ij} f_{;k} \quad \text{and}
$$
$$
\mu\{f_{;k}(\rho_\varepsilon)_{ij} - f_{;j}(\rho_\varepsilon)_{ik}\} = \mu(m-1)\{f_{;j}\varepsilon_{ij} - \varepsilon_{ik} f_{;k}\}.
$$

Since $\mu(m-1) \ne -1$, this implies $f_{;k} = 0$ for all k. We may rescale f to ensure that $f = \mathbb{1}$. Since $\mathcal{H}\mathbb{1} = 0$, this implies $\mu\rho_\varepsilon = 0$. This shows that $\mu = 0$. $\square$

We have the following consequence.

Corollary 14.28 *Let $S^2 := \{(x, y, z) \in \mathbb{R}^3 : x^2 + y^2 + z^2 = 1\}$ be given the metric inherited from the flat Euclidean metric. Then S^2 has no affine gradient Ricci solitons,*

$$\mathcal{Y}(S^2) = \mathbb{1} \cdot \mathbb{R}, \ \mathcal{Q}(S^2) = \mathrm{span}\{x, y, z\}, \ and \ E(\mu, S^2) = \{0\} \ for \ \mu \neq 0 \ and \ \mu \neq -1 \,.$$

Proof. Take $\varepsilon = \varepsilon_S$ as defined in Equation (14.3.a) and apply Lemma 14.26 and Lemma 14.27. □

14.3.4 THE HYPERBOLIC PLANE. Let $ds^2 = (x^1)^{-2}((dx^1)^2 + (dx^2)^2)$. A direct computation using the Koszul formula (see Theorem 3.7 of Book I) shows that the non-zero Christoffel symbols of the Levi-Civita connection are

$$\Gamma_{11}{}^1 = \Gamma_{12}{}^2 = -(x^1)^{-1} \quad \text{and} \quad \Gamma_{22}{}^1 = (x^1)^{-1} \,.$$

This is the geometry $\mathcal{N}_4^3$ of Definition 14.1. Thus, in particular, the Ricci tensor is given by $\rho = -(x^1)^{-2}((dx^1)^2 + (dx^2)^2)$. This surface is Einstein with Einstein constant -1. We summarize briefly well known facts concerning hyperbolic geometry as an introduction to our discussion of the Lorentzian analogue in Section 14.3.6. Set $z := x^2 + \sqrt{-1}x^1$ to identify $\mathbb{R}^+ \times \mathbb{R}$ with the upper half-plane $\mathbb{H} := \{z \in \mathbb{C} : \Im(z) > 0\}$. Let $\mathrm{SL}(2, \mathbb{R})$ be the Lie group of linear transformations of $\mathbb{R}^2$ with determinant 1. We refer to Lemma 6.25 of Book II for further information concerning this Lie group. The subgroup of diagonal matrices $\pm \mathrm{Id}$ is a normal subgroup of $\mathrm{SL}(2, \mathbb{R})$ and we set $\mathrm{PSL}(2, \mathbb{R}) := \mathrm{SL}(2, \mathbb{R})/\{\pm \mathrm{Id}\}$. We define a *linear fractional transformation* preserving $\mathbb{H}$ by setting:

$$T_A(z) := \frac{az + b}{cz + d} \quad \text{for} \quad A = \begin{pmatrix} a & b \\ c & d \end{pmatrix} \in \mathrm{SL}(2, \mathbb{R}) \,.$$

The following is well known. Since it follows from a direct computation, we omit the proof.

Lemma 14.29 *Let A and B belong to $\mathrm{SL}(2, \mathbb{R})$. Then $T_A \circ T_B = T_{AB}$. We have T_A is an orientation-preserving isometry of $\mathbb{H}$, and the map $A \to T_A$ identifies $\mathrm{PSL}(2, \mathbb{R})$ with the group of orientation-preserving isometries of $\mathbb{H}$.*

The curve $\sigma(t) = (e^t, c)$ satisfies the geodesic equation $x^1 \ddot{x}^1(t) - \dot{x}^1(t)\dot{x}^1(t) = 0$. Thus horizontal rays from the vertical axis are complete affine geodesics. Linear fractional transformations preserve angles and map straight lines/circles to straight lines/circles. It now follows that any circle whose center is on the vertical axis (i.e., the imaginary axis) is a complete affine geodesic and thus $\mathbb{H}$ is geodesically complete.

14.3.5 THE HYPERBOLIC PSEUDO-SPHERE $\mathfrak{H}$. Define a non-degenerate inner product of signature $(1, 2)$ on $\mathbb{R}^3$ by setting $\langle x, y \rangle = x^1 y^1 + x^2 y^2 - x^3 y^3$. Give the level set

$$\mathfrak{H} := \{ \vec{x} : \langle \vec{x}, \vec{x} \rangle = -1 \text{ and } x^3 > 0 \}$$

the induced Riemannian inner product to define the hyperbolic pseudo-sphere $\mathfrak{H}$. Subsequently in Section 14.3.7, we will consider the level set $\langle \vec{x}, \vec{x} \rangle = +1$ in studying the Lorentzian hyperbolic plane. We assume $x^3 > 0$ to ensure $\mathfrak{H}$ is connected as the equation $\langle \vec{x}, \vec{x} \rangle = -1$ is a hyperboloid of two sheets. The x^3 axis is vertical.

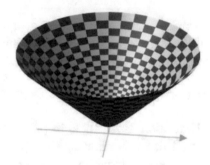

Lemma 14.30

1. *The Lorentz group $O(1, 2)$ acts transitively on $\mathfrak{H}$ by isometries. The level set $\mathfrak{H}$ is geodesically complete, $\rho = -g$, $\nabla \rho = 0$, and $\mathcal{Q}(\mathfrak{H}) = \operatorname{span}\{x^1, x^2, x^3\}$.*

2. *The exponential map is surjective.*

3. *Let $\Phi(x^1, x^2) := \frac{1}{2x^1}((x^1)^2 + (x^2)^2 - 1, \quad 2x^2, \quad (x^1)^2 + (x^2)^2 + 1)$. Then Φ is an affine diffeomorphism between $\mathbb{H}$ and $\mathfrak{H}$.*

Proof. Adopt the notation of Equation (14.3.a) and set

$$\varepsilon_{\mathfrak{H}} := \begin{pmatrix} -1 & 0 & 0 \\ 0 & -1 & 0 \\ 0 & 0 & +1 \end{pmatrix}.$$

Then $\varepsilon_{\mathcal{H}} = -\langle \cdot, \cdot \rangle$ so $g = -g_{\varepsilon_{\mathcal{H}}}$ but allowing for the changes in sign, Assertion 1 follows from Lemma 14.26. Let $\sigma(t) := (u \sinh(t), v \sinh(t), \cosh(t))$. The analysis of Lemma 14.29 shows that these curves are affine geodesics. Consequently, the exponential map is surjective. Let $\Phi = (\Phi_1, \Phi_2, \Phi_3)$. We verify $\Phi_1^2 + \Phi_2^2 - \Phi_3^2 = \langle \Phi, \Phi \rangle = -1$ so Φ takes values in $\mathfrak{H}$. We can recover x^1 from $\Phi_1 + \Phi_3$, and once x^1 is known, we can recover x^2 from Φ_2. Thus Φ is a smooth embedding. We use Lemma 14.4 to see $\mathcal{Q}(\mathbb{H}) = \operatorname{span}\{\frac{1}{x^1}, \frac{x^2}{x^1}, \frac{(x^1)^2 + (x^2)^2}{x^1}\}$. An algebraic computation shows $\mathcal{Q}(\mathbb{H}) = \operatorname{span}\{\Phi_1, \Phi_2, \Phi_3\}$ so $\mathcal{Q}(\mathbb{H}) = \Phi^* \operatorname{span}\{x^1, x^2, x^3\}$. By Assertion 1, $\mathcal{Q}(\mathfrak{H}) = \operatorname{span}\{x^1, x^2, x^3\}$. Theorem 12.34 then implies Φ is an affine map. As the exponential map is a diffeomorphism for both $\mathbb{H}$ and $\mathfrak{H}$, Φ is a diffeomorphism. $\square$

Let Φ be the map of Lemma 14.30. Since Φ is an affine map, Φ intertwines the Ricci tensors. Since the metrics are essentially given by the Ricci tensors up to sign, Φ is an isometry. One can also check this directly by verifying that

$$\Phi^*(g_{\mathfrak{H}}) = d\Phi_1^2 + d\Phi_2^2 - d\Phi_3^2 = (x^1)^{-2}((dx^1)^2 + (dx^2)^2).$$

14.3.6 THE LORENTZIAN HYPERBOLIC PLANE. The Euclidean inner product is defined by the quadratic form $((x^1, x^2), (y^1, y^2)) = x^1 y^1 + x^2 y^2$. The Lorentzian inner product is defined by setting $\langle (x^1, x^2), (y^1, y^2) \rangle := -x^1 y^1 + x^2 y^2$. The null vectors are defined by the equation $x^1 = \pm x^2$. In analogy with the hyperbolic plane, let $\mathbb{L} = (\mathbb{R}^+ \times \mathbb{R})$ with the Lorentzian metric $ds^2 = (x^1)^{-2}\{-(dx^1)^2 + (dx^2)^2\}$. The Koszul formula now yields

$$\Gamma_{11}{}^1 = \Gamma_{12}{}^2 = \Gamma_{22}{}^1 = -(x^1)^{-1}.$$

This is the structure $\mathcal{N}_3^3$ of Definition 14.1 and thus by Lemma 14.3,

$$\rho = (x^1)^{-2}(-(dx^1)^2 + (dx^2)^2),$$

so the Einstein constant is 1. We remark that we could equally well have taken

$$ds^2 = (x^1)^{-2}\{(dx^1)^2 - (dx^2)^2\}$$

since we are only interested in the underlying affine structure and the resulting Levi-Civita connection is the same; the Einstein constant here would be -1. Thus these two structures are not isometric, although the underlying affine structure is the same. As was the case for the hyperbolic plane, the linear fractional transformations play a central role. Let

$$\mathrm{Id} := \begin{pmatrix} 1 & 0 \\ 0 & 1 \end{pmatrix} \quad \text{and} \quad \iota := \begin{pmatrix} 0 & 1 \\ 1 & 0 \end{pmatrix}.$$

Let $\hat{\mathbb{C}} := \mathrm{span}_{\mathbb{R}}\{\mathrm{Id}, \iota\} \subset M_2(\mathbb{R})$ be the *para-complex* numbers; we refer to Cruceanu, Fortuny, and Gadea [20] for further details. Since $\iota^2 = \mathrm{Id}$, $\hat{\mathbb{C}}$ is a Abelian 2-dimensional unital algebra; unlike the complex numbers, $\hat{\mathbb{C}}$ is not a field as there are zero divisors. We set $\hat{z} = x^2\,\mathrm{Id} + x^1\iota$. Let $\hat{z}^* = x^2\,\mathrm{Id} - x^1\iota$. We compute $\hat{z}\,\hat{z}^* = (x^2)^2 - (x^1)^2$. Consequently, $\hat{z}$ is invertible if and only if $x^1 \neq \pm x^2$, i.e., $\hat{z}$ is not a null vector. Let

$$\hat{T}_A(\hat{z}) = \frac{a\hat{z} + b}{c\hat{z} + d} \quad \text{for} \quad A = \begin{pmatrix} a & b \\ c & d \end{pmatrix} \in \mathrm{SL}(2, \mathbb{R}).$$

The transformation $\hat{T}_A$ is a *para-complex linear fractional transformation*. Unlike the complex setting, $\hat{T}_A$ is not defined for $c\hat{z} + d = 0$, but only on the open dense subset where we have that $cx^2 + d \neq \pm cx^1$. The following result is due to Catoni et al. [16] and provides the appropriate generalization of Lemma 14.29 to this setting; as it follows by a direct computation, we shall omit the proof.

Lemma 14.31   *Let A and B belong to $\mathrm{SL}(2,\mathbb{R})$. Then $\hat{T}_A \circ \hat{T}_B = \hat{T}_{AB}$. We have $\hat{T}_A$ is an orientation-preserving isometry of $\mathbb{L}$, where defined, and the map $A \to \hat{T}_A$ identifies $\mathrm{PSL}(2,\mathbb{R})$ with the group of orientation-preserving isometries of $\mathbb{L}$.*

The Lorentzian hyperbolic plane is the only non-complete symmetric space of Type $\mathcal{B}$, and the exponential map is not surjective, although it is 1-1. Let $X = \xi_1 \partial_{x^1} + \xi_2 \partial_{x^2}$; X is a *null vector field* if $\xi_1 = \pm \xi_2$, X is a *timelike vector field* if $|\xi_1| > |\xi_2|$, and X is a *spacelike vector field* if $|\xi_1| < |\xi_2|$. The following result is due to D'Ascanio, Gilkey, and Pisani [22]; we shall omit the proof in the interests of brevity.

Theorem 14.32

1. *The affine geodesics of $\mathbb{L}$ have one of the following forms for some $\alpha, \beta, c \in \mathbb{R}$, up to rescaling:*

   (a) $\sigma(t) = (e^t, \alpha)$ for $-\infty < t < \infty$. This affine geodesic is complete.

   (b) $\sigma(t) = (t^{-1}, \pm t^{-1} + \alpha)$ for $0 < t < \infty$. This affine geodesic is incomplete at one end and complete at the other end.

   (c) $\sigma(t) = (\frac{1}{c \sinh(t)}, \pm \frac{\coth(t)}{c} + \beta)$ for $t \in (0, \infty)$ and $c > 0$. This tends asymptotically to the line $x^1 = 0$ as $t \to \infty$ and escapes to the right as $t \to 0$. These affine geodesics are incomplete at one end and complete at the other. These affine geodesics all have infinite (and negative) length.

   (d) $\sigma(t) = (\frac{1}{c \sin(t)}, \pm \frac{\cot(t)}{c} + \beta)$ for $t \in (0, \pi)$ and $c > 0$. These affine geodesic escape upwards and to the right as $t \to 0$ and downwards and to the right as $t \to \pi$. The affine geodesic σ is incomplete at both ends and has total length π.

   (e) The affine geodesics in (c) and (d) solve the equation $(x^1)^2 - \frac{\lambda}{c^2} = (x^2 + \beta)^2$; they are hyperbolas; the affine geodesic is "vertical" if $\lambda = +1$, "horizontal" if $\lambda = -1$, and null if $\lambda = 0$.

2. *The exponential map is an embedding of $T_P M$ which omits a region of $\mathbb{L}$.*

We picture the geodesic structure below; the region omitted by the exponential map consists of the regions $x^2 \geq 1 + x^1$ and $x^2 \leq -1 - x^1$. The "vertical" affine geodesics (in red) point up (resp. below) and to the right; they lie above and below the two null half-lines with slope $\pm \frac{\pi}{4}$. The "horizontal" affine geodesics (in black) point to the right and the left and lie between the null half-lines with slope $\pm \frac{\pi}{4}$. The x^1 axis is horizontal and the x^2 axis is vertical.

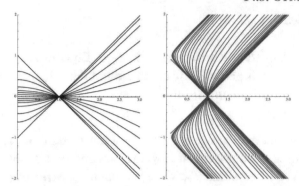

14.3.7 THE LORENTZIAN PSEUDO-SPHERE $\mathfrak{L}$.

We continue our investigation of the Lorentzian hyperbolic plane using a different model following the discussion of D'Ascanio, Gilkey, and Pisani [22]. As before, we set $\langle x, y \rangle = x^1 y^1 + x^2 y^2 - x^3 y^3$. We now consider the level set $\{\vec{x} : \langle \vec{x}, \vec{x} \rangle = +1\}$; the induced metric is a Lorentzian inner product defining the Lorentzian pseudo-sphere $\mathfrak{L}$. The x^3 axis is vertical.

The pseudo-sphere $\mathfrak{L}$ is not simply connected since the hyperboloid of one sheet given by $(x^1)^2 + (x^2)^2 - (x^3)^2 = 1$ is diffeomorphic to $S^1 \times \mathbb{R}$. Set

$$T(x^1, x^2) := (\cosh(x^1)\cos(x^2), \cosh(x^1)\sin(x^2), \sinh(x^1)) ;$$

T exhibits $\mathbb{R}^2$ as the universal cover $\tilde{\mathfrak{L}}$ of $\mathfrak{L}$.

Lemma 14.33

1. $\mathfrak{L}$ is geodesically complete and $\mathcal{Q}(\mathfrak{L}) = \mathrm{span}\{x^1, x^2, x^3\}$.

2. The exponential map in $\mathfrak{L}$ is not surjective.

3. $\mathcal{Q}(\tilde{\mathfrak{L}}) = \mathrm{span}\{\cosh(x^1)\cos(x^2), \cosh(x^1)\sin(x^2), \sinh(x^1)\}$, $\Gamma_{22}{}^2 = \cosh(x^1)\sinh(x^1)$ and $\Gamma_{12}{}^2 = \frac{\sinh(x^1)}{\cosh(x^1)}$ give the induced affine structure on $\mathbb{R}^2$, $\rho_{\tilde{\mathfrak{L}}} = \mathrm{diag}(-1, \cosh^2(x^1))$, and $\nabla \rho_{\tilde{\mathfrak{L}}} = 0$.

4. Let $\Phi(x^1, x^2) := (2x^1)^{-1}(1 + (x^1)^2 - (x^2)^2, \quad 2x^2, \quad -1 + (x^1)^2 - (x^2)^2)$. Then Φ is a

proper affine embedding of $\mathbb{L}$ *in* $\mathfrak{L}$.

Proof. We have $\langle \cdot, \cdot \rangle$ is defined by the bilinear form $\varepsilon_{\mathfrak{L}}$ of Equation (14.3.a). We may therefore apply Lemma 14.29 to see $\mathfrak{L}$ is geodesically complete and that $O(2,1)$ acts transitively on

$$\mathfrak{L} \cup -\mathfrak{L}.$$

To see that the exponential map is not surjective, we must examine the geodesic structure in a bit more detail. Since $\mathfrak{L}$ is homogeneous, we may assume that $e_1 = (1,0,0)$ in proving the remaining assertions. We observe that $T_{e_1}\mathfrak{L} = e_1^{\perp} = \mathrm{span}\{e_2, e_3\}$. Let $\xi = ae_2 + be_3$.

Case 1. Assume $a^2 - b^2 > 0$ so ξ is spacelike. We rescale ξ to ensure that $a^2 - b^2 = 1$. Let $\sigma(\theta) = \cos(\theta)e_1 + \sin(\theta)\xi : \mathbb{R} \to \mathfrak{L}$. Since $\ddot{\sigma} = -\sigma$, $\ddot{\sigma} \perp T_{\sigma}\mathfrak{L}$. Consequently, $\ddot{\sigma} \perp \mathfrak{L}$. This implies σ is an affine geodesic which is defined for all time.

Case 2. Assume $a^2 - b^2 = 0$ so ξ is null. We can let $\sigma(t) = e_1 + t\xi$. Since $\ddot{\sigma} = 0$, this is an affine geodesic which extends for all time.

Case 3. Assume $a^2 - b^2 < 0$ so ξ is timelike.

We can rescale ξ to ensure that $b^2 - a^2 = 1$. Let $\sigma(t) = \cosh(t)e_1 + \sinh(t)\xi$. Again, $\ddot{\sigma} \perp \mathfrak{L}$ so this affine geodesic is defined for all time.

We note that affine geodesics can never reach $-e_1 + \xi$ for $0 \neq \xi$ null, and hence the exponential map is not surjective. We prove the third assertion by computing:

$$\partial_{x^1} T = (\cos(x^2)\sinh(x^1), \sin(x^2)\sinh(x^1), \cosh(x^1)),$$
$$\partial_{x^2} T = (-\sin(x^2)\cosh(x^1), \cos(x^2)\cosh(x^1), 0),$$
$$g_{11} = \langle \partial_{x^1} T, \partial_{x^1} T \rangle = -1, \qquad g_{12} = \langle \partial_{x^1} T, \partial_{x^2} T \rangle = 0,$$
$$g_{22} = \langle \partial_{x^2} T, \partial_{x^2} T \rangle = \cosh^2(u).$$

We use the Koszul formula (see Theorem 3.7 of Book I) to see $\Gamma_{122} = \cosh(x^1)\sinh(x^1)$ and $\Gamma_{221} = -\cosh(x^1)\sinh(x^2)$. We raise indices to determine $\Gamma_{12}{}^2$ and $\Gamma_{22}{}^1$. We perform a direct computation to find $\rho_{\tilde{\mathfrak{L}}}$ and $\nabla\rho_{\tilde{\mathfrak{L}}}$; we use Lemma 14.29 to see $\mathcal{Q}(\mathfrak{L}) = \mathrm{span}\{x^1, x^2, x^3\}$ and then pull-back to compute $\mathcal{Q}(\tilde{\mathfrak{L}})$.

We compute $\Phi_1^2 + \Phi_2^2 - \Phi_3^2 = 1$. Consequently, Φ takes values in $\mathfrak{L}$. We can recover x^1 from $\Phi_1 - \Phi_3$ and once x^1 is known, we can recover x^2 from Φ_2. Thus Φ is a smooth embedding; since $\mathfrak{L}$ is not simply connected and $\mathbb{L}$ is simply connected, the embedding is proper. By Lemma 14.4,

$$\mathcal{Q}(\mathbb{L}) = \mathrm{span}\{\frac{1}{x^1}, \frac{x^2}{x^1}, \frac{-(x^1)^2 + (x^2)^2}{x^1}\}.$$

By Assertion 1, the coordinate functions $\{x^1, x^2, x^3\}$ span $\mathcal{Q}(\mathfrak{L})$. Let $\Phi = (\Phi_1, \Phi_2, \Phi_3)$. We observe that $\mathcal{Q}(\mathbb{L}) = \mathrm{span}\{\Phi_1, \Phi_2, \Phi_3\}$. Since $\Phi^*(\mathcal{Q}(\mathfrak{L})) = \mathcal{Q}(\mathbb{L})$, Theorem 12.34 yields that Φ is an affine map. $\qquad \square$

We have the following picture of the geodesic structure in $\tilde{\mathfrak{L}}$; the corresponding geodesic structure in $\mathfrak{L}$ may be obtained by identifying (x^1, x^2) with $(x^1, x^2 + 2n\pi)$ – i.e., wrapping it up to form a cylinder. The x^1 axis is horizontal.

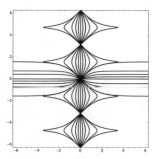

Let Φ be the map of Lemma 14.33. We note as before that since Φ is an affine map, Φ preserves the Ricci tensor, and hence the metric so Φ is an isometry. One can also compute directly $d\Phi_1^2 + d\Phi_2^2 - d\Phi_3^2 = (x^1)^{-2}(-(dx^1)^2 + (dx^2)^2)$.

CHAPTER 15

Applications of Affine Surface Theory

In Section 15.1, we present some basic material concerning the Riemannian extension and other notions which we shall need. In Section 15.2, we discuss a family of VSI (vanishing scalar invariants) manifolds. In Section 15.3, we present material of Calviño-Louzao et al. [14] concerning Bach flat manifolds.

15.1 PRELIMINARY MATTERS

In Section 15.1.1, we define the Bach, Cotton, and Weyl tensors and we introduce Walker coordinates. In Section 15.1.2, we discuss nilpotent endomorphisms of the tangent bundle of a surface. In Section 15.1.3, we discuss various modifications of the Riemannian extension which provide useful metrics on the cotangent bundle. In Section 15.1.4, we construct Bach flat structures on the cotangent bundle of an affine surface. We introduce quasi-Einstein geometry in Section 15.1.5.

15.1.1 BASIC DEFINITIONS. Let $\mathcal{N} = (N, g)$ be a pseudo-Riemannian manifold of dimension $n \geq 4$. Let ∇^g be the associated Levi-Civita connection, and let $R(\cdot, \cdot)$ be the curvature operator of ∇^g. Let

$$R_{ijk\ell} := g(R(\partial_{x^i}, \partial_{x^j})\partial_{x^k}, \partial_{x^\ell})$$

be the components of the curvature tensor. Let $\rho_{ij} = R_{kij}{}^k$ be the components of the Ricci tensor, and let $\tau := g^{ij}\rho_{ij}$ be the scalar curvature (see Section 3.5.2 of Book I). Let Δ be the scalar Laplacian. The *Weyl tensor* is conformally invariant. It is defined by setting

$$W_{ijk\ell} := R_{ijk\ell} + \frac{\rho_{i\ell}g_{jk} - \rho_{ik}g_{j\ell} + R_{jk}g_{i\ell} - R_{j\ell}g_{ik}}{n-2} + \tau\frac{g_{ik}g_{j\ell} - g_{i\ell}g_{jk}}{(n-1)(n-2)}.$$

We say that $\mathcal{N}$ is *conformally flat* if $W = 0$ or, equivalently, if there exists a coordinate atlas so $ds^2 = e^\phi((dx^1)^2 + \cdots + (dx^n)^2)$. Suppose that $n = 4$ and that N is oriented. We may decompose W under the eigenvalues of the Hodge $\star$ operator (see Section 5.2 of Book II) into components $W^\pm$. We say that $\mathcal{N}$ is self-dual (resp. anti-self-dual) if W^+ (resp. W^-) vanishes. One says that $\mathcal{N}$ is *half conformally flat* if either $W^+ = 0$ or $W^- = 0$ but $\mathcal{N}$ is not conformally flat; $\mathcal{N}$ is conformally flat if and only if $W = 0$, i.e., $W^+ = 0$ and $W^- = 0$.

The *Cotton tensor* is conformally invariant in dimension 3 and vanishes if $\mathcal{M}$ is conformally flat; it is defined in arbitrary dimension by setting

$$\mathfrak{C}(X, Y, Z) := (\nabla_X \rho)(Y, Z) - (\nabla_Y \rho)(X, Z) - \frac{d\tau(X)g(Y, Z) - d\tau(Y)g(X, Z)}{2(n-1)}.$$

The *Bach tensor* of $\mathcal{N}$ is defined by setting:

$$\mathfrak{B}_{ij} = \nabla^k \nabla^\ell W_{kij\ell} + \tfrac{1}{2}\rho^{k\ell} W_{kij\ell}.$$

It is a conformally invariant tensor. We say that $\mathcal{N}$ is *Bach flat* if $\mathfrak{B} = 0$. We will construct various examples of Bach flat manifolds in Theorem 15.4 which are not conformally flat, i.e., which have non-vanishing Weyl tensors.

We say that a subbundle $\mathcal{D}$ of the tangent bundle of a pseudo-Riemannian manifold $\mathcal{N} = (N, g)$ is *totally isotropic* if the restriction of the metric g to $\mathcal{D}$ is trivial. If $\dim(N) = 2\ell$ is even, or if $\dim(N) = 2\ell + 1$ is odd and if $\mathcal{D}$ is totally isotropic, then $\dim(\mathcal{D}) \leq \ell$; we say that $\dim(\mathcal{D})$ is *maximal* if equality holds. The bundle is said to be *parallel* if any section ξ to $\mathcal{D}$ satisfies $\nabla_X \xi$ is again a section to $\mathcal{D}$ for any tangent vector field X. We say that $\mathcal{N} = (N, g)$ is a *Walker manifold* if there is a non-trivial totally isotropic parallel subbundle $\mathcal{D}$ of the tangent bundle of maximal dimension; $\mathcal{N}$ is said to be *strict* if $\mathcal{D}$ is generated by parallel vector fields. If $n = 4$ and if $\mathcal{N}$ is a Walker manifold, then $\mathcal{N}$ has neutral signature. The work of Walker [61](see also the discussion of Section 5.3 of Brozos-Vázquez et al. [8]) when restricted to $n = 4$ yields the following result:

Theorem 15.1 *A pseudo–Riemannian manifold $\mathcal{N}$ of dimension 4 is said to be a Walker manifold if and only if there is a coordinate atlas with local coordinates (x^1, x^2, y_1, y_2) and if there exist smooth functions $a(\cdot), b(\cdot),$ and $c(\cdot)$ so that*

$$\begin{aligned} g \; = \; & dx^1 \otimes dy_1 + dy_1 \otimes dx^1 + dx^2 \otimes dy_2 + dy_2 \otimes dx^2 \\ & + a(x^1, x^2, y_1, y_2)dx^1 \otimes dx^1 + b(x^1, x^2, y_1, y_2)dx^2 \otimes dx^2 \\ & + c(x^1, x^2, y_1, y_2)(dx^1 \otimes dx^2 + dx^2 \otimes dx^1). \end{aligned}$$

In this setting, the distinguished distribution takes the form $\mathcal{D} = \mathrm{span}\{\partial_{y_1}, \partial_{y_2}\}$. Such a manifold N has a distinguished orientation given locally by $dx^1 \wedge dx^2 \wedge dy_1 \wedge dy_2$; this ensures that $\mathcal{D}$ is self-dual. Thus self-duality and anti-self-duality are not equivalent concepts in this context. We refer to Brozos-Vázquez et al. [8] for further details.

15.1.2 NILPOTENT ENDOMORPHISMS.

Let T be an endomorphism of the tangent bundle of a manifold M. Let $T\partial_{x^j} = T^i{}_j \partial_{x^i}$ define the components of T. We then have that $T = T^i{}_j \partial_{x^i} \otimes dx^j$; consequently, T is said to be of *Type* $(1, 1)$. We say that T is *nilpotent* if $T^2 = 0$; nilpotent endomorphism will play a central role in our discussion of Bach flat manifolds in Section 15.3. The following is a useful technical observation.

Lemma 15.2   *Let $0 \neq T$ be a nilpotent tensor of Type (1,1) on a surface M. We can choose local coordinates near any point of M so that $T = \partial_{x^1} \otimes dx^2$, i.e., so $T^1{}_2$ is the only non-zero component of T.*

Proof.   Let $0 \neq T$ be nilpotent. Fix a point P of M. We work locally in a small neighborhood of P. Choose a non-zero vector field W_2 defined near P so that $W_1 := TW_2 \neq 0$. Choose local coordinates (y^1, y^2) on M centered at P so that $W_1 = \partial_{y^1}$. Since T is nilpotent, $T\partial_{y^1} = T(TW_2) = 0$. Since T is nilpotent, range$\{T\}$ = ker$\{T\}$. Since ∂_{y^1} spans range$\{T\}$, we have that $T\partial_{y^2} = f\partial_{y^1}$ for some non-zero function f. Set $\tilde{W}_1 = f\partial_{y^1}$ and $\tilde{W}_2 = F\partial_{y^1} + \partial_{y^2}$ where F remains to be determined. We compute

$$[\tilde{W}_1, \tilde{W}_2] = [f\partial_{y^1}, F\partial_{y^1} + \partial_{y^2}] = \{f\partial_{y^1}F - F\partial_{y^1}f - \partial_{y^2}f\}\partial_{y^1}.$$

The equation $[\tilde{W}_1, \tilde{W}_2] = 0$ is equivalent to the ODE $\partial_{y^1}F = f^{-1}\{F\partial_{y^1}f + \partial_{y^2}f\}$. We solve this ODE with initial condition $F(0, y^2) = 0$ to ensure that $[\tilde{W}_1, \tilde{W}_2] = 0$. We then use the *Frobenius Theorem* to choose local coordinates $\{x^1, x^2\}$ so $\tilde{W}_1 = \partial_{x^1}$ and $\tilde{W}_2 = \partial_{x^2}$. We then have $T\partial_{x^1} = \partial_{x^2}$ and $T\partial_{x^2} = 0$ so $T = \partial_{x^1} \otimes dx^2$ as desired.   □

Ferdinand Georg Frobenius (1849–1917)

15.1.3 METRICS ON THE COTANGENT BUNDLE.
The subject arises first in the classic papers of Patterson and Walker [56] and Walker [61]. We also refer to related work of Afifi [1] and of Derdzinski and Roter [26]. Let $\mathcal{M} = (M, \nabla)$ be an affine manifold. Definition 11.39 of Book III gives various canonically defined metrics on the cotangent bundle of M. We recall these metrics for the convenience of the reader. Choose local coordinates (x^1, x^2) on M. We introduce *dual coordinates* (y_1, y_2) on the cotangent bundle T^*M by expressing any 1-form locally in the form $\omega = y_1 dx^1 + y_2 dx^2$. The classical *Riemannian extension* is defined by setting:

$$g_\nabla = dx^i \otimes dy_i + dy_i \otimes dx^i - 2y_k\Gamma_{ij}{}^k(\vec{x})dx^i \otimes dx^j. \tag{15.1.a}$$

If $\Phi = \Phi_{ij}dx^i \otimes dx^j$ is a smooth symmetric 2-tensor on M, then we can use Φ to deform the metric g_∇ of Equation (15.1.a) by setting:

$$g_{\nabla,\Phi} = g_\nabla + \Phi_{ij}(\vec{x})dx^i \otimes dx^j. \tag{15.1.b}$$

We omit the proof of the following result as it is a direct computation.

Lemma 15.3   *Let $g = g_{\nabla,\Phi}$ and let π be the canonical projection from the cotangent bundle to M. Then $\mathcal{H}_g(\pi^* f) = \pi^* \mathcal{H}(f)$, $\|d\pi^* f\|_g^2 = 0$, and $\rho_g = 2\pi^* \rho_{s,\nabla}$.*

Finally, suppose that $X = a^i \partial_{x^i}$ is a vector field on M and that $S = S^j{}_i$ and $T = T^j{}_i$ are tensors of Type (1,1). We modify $g_{\nabla,\Phi}$ to define

$$g_{\nabla,\Phi,T,S,X} = g_{\nabla,\Phi} + y_r y_s T^r{}_i(\vec{x}) S^s{}_j(\vec{x}) dx^i \otimes dx^j + a^i y_i y_j y_k dx^j \otimes dx^k . \qquad (15.1.c)$$

By Lemma 11.40 of Book III, these metrics are invariantly defined and independent of the particular local coordinate system (x^1, x^2) chosen; it is immediate from Theorem 15.1 that these all are Walker metrics. If one or more of these parameters is zero, we shall drop it from the notation.

15.1.4 BACH FLAT GEOMETRY. We take $X = 0$ and $S = T$ in Equation (15.1.c) to define the metric $g_{\nabla,\Phi,T}$. We use the metrics $g_{\nabla,\Phi,T}$ to provide examples of Bach flat neutral signature manifolds; we will consider these geometries in more detail in Section 15.3 and for the moment content ourselves with constructing some non-trivial examples.

Theorem 15.4   *Let $\mathcal{M} = (M, \nabla)$ be an affine surface, let T be a tensor of Type (1, 1), and let Φ be a symmetric 2-tensor. Let $\mathcal{N} := (T^* M, g_{\nabla,\Phi,T})$.*

1. *Suppose that $T(P) = \lambda(P) \operatorname{Id}$ is a scalar multiple of the identity for every point P of M. Then $\mathcal{N}$ is half conformally flat, and hence Bach flat.*

2. *Suppose that $T(P)$ is not a scalar multiple of the identity for any point of M. If $\mathcal{N}$ is Bach flat, then T is nilpotent.*

3. *Suppose that T is nilpotent and non-vanishing on M. Fix $P \in M$ and use Lemma 15.2 to choose local coordinates on an open neighborhood $\mathcal{O}$ of P so that $T = \partial_{x^1} \otimes dx^2$. The following assertions are equivalent on $\mathcal{O}$:*

   (a) $\mathcal{N}$ is Bach flat.

   (b) $\Gamma_{11}{}^2 = 0$ and $(\Gamma_{11}{}^1)^2 - \Gamma_{11}{}^1 \Gamma_{12}{}^2 + \partial_{x^1}(\Gamma_{11}{}^1 - \Gamma_{12}{}^2) = 0$.

We note that the auxiliary tensor Φ plays no role in the analysis. We can express the conditions of Assertion 3-b in the form

$$\Gamma_{11}{}^2 = 0, \qquad \Gamma_{11}{}^1 = -\phi^{(1,0)}, \qquad \Gamma_{12}{}^2 = \Gamma_{11}{}^1 + c \cdot e^\phi$$

for smooth functions $c = c(x^2)$ and $\phi = \phi(x^1, x^2)$. Since we are interested in Bach flat manifolds which are not half conformally flat, we will concentrate on the case in which T is nilpotent in Section 15.3.

Proof. A direct computation shows that if $T = f\,\mathrm{Id}$ for $f \in C^\infty(M)$, then $\mathcal{N}$ is half conformally flat and that $\mathfrak{B} = 0$; this establishes Assertion 1 of Theorem 15.4.

We now turn to the proof of Assertion 2. Let $\Theta_{ijk\ell}$ be the coefficient of $y_i\,y_j$ in $\mathfrak{B}_{k\ell}$. A straightforward computation shows that $\Theta_{ijk\ell}$ is a polynomial which is homogeneous of degree 6 in the $T^u{}_v$ variables for $k, \ell \in \{1, 2\}$ and zero otherwise; the Christoffel symbols and their derivatives, the auxiliary endomorphism Φ and its derivatives, and the derivatives of T do not appear in these terms. Consequently, $\Theta = \{\Theta_{ijk\ell}\}$ is tensorial. Assume that $\mathcal{N}$ is Bach flat. This implies $\Theta(T) = 0$. We suppose $T(P)$ is not a scalar multiple of the identity. Let $\{\lambda_1, \lambda_2\}$ be the (possibly complex) eigenvalues of $T(P)$. We can make a complex linear change of coordinates in the $\{x^1, x^2\}$ variables to put T in upper triangular form; this induces a corresponding dual complex linear change of coordinates in the $\{y_1, y_2\}$ variables. This is, of course, just Jordan normal form. Thus we may assume that:

$$T(P) := \begin{pmatrix} \lambda_1 & \varepsilon \\ 0 & \lambda_2 \end{pmatrix}.$$

Since $T(P)$ is not diagonal, we have either that $\lambda_1 \neq \lambda_2$ or $\lambda_1 = \lambda_2$ but $\varepsilon \neq 0$. We distinguish cases.

Case 1. Suppose $\lambda_1 \neq \lambda_2$. We compute that

$$\Theta_{1111}(T(P)) = \tfrac{1}{6}\lambda_1^2(\lambda_1 - \lambda_2)^2(\lambda_1^2 + \lambda_1\lambda_2 - 5\lambda_2^2),$$
$$\Theta_{2222}(T(P)) = \tfrac{1}{6}\lambda_2^2(\lambda_1 - \lambda_2)^2(-5\lambda_1^2 + \lambda_1\lambda_2 + \lambda_2^2).$$

Note that the parameter ε does not appear; these two terms are not sensitive to the precise Jordan normal form but only to the eigenvalues. Since $\mathfrak{B} = 0$ and since $\lambda_1 - \lambda_2 \neq 0$, we obtain

$$\lambda_1^2(\lambda_1^2 + \lambda_1\lambda_2 - 5\lambda_2^2) = 0 \quad \text{and} \quad \lambda_2^2(-5\lambda_1^2 + \lambda_1\lambda_2 + \lambda_2^2) = 0. \tag{15.1.d}$$

If $\lambda_2 = 0$, then $\lambda_1 \neq 0$ and the first identity of Equation (15.1.d) fails. Similarly, if $\lambda_1 = 0$, then $\lambda_2 \neq 0$ and the second identity of Equation (15.1.d) fails. We may conclude therefore that $\lambda_1 \neq 0$ and $\lambda_2 \neq 0$. We now obtain

$$\lambda_1^2 + \lambda_1\lambda_2 - 5\lambda_2^2 = 0 \quad \text{and} \quad -5\lambda_1^2 + \lambda_1\lambda_2 + \lambda_2^2 = 0.$$

Subtracting these two identities yields $6\lambda_1^2 - 6\lambda_2^2 = 0$. Since $\lambda_1 \neq \lambda_2$, we have $\lambda_1 = -\lambda_2$, so $-5\lambda_1^2 = 0$ and again $\lambda_1 = \lambda_2 = 0$ which is false.

Case 2. Suppose that $\lambda_1 = \lambda_2$ but $\varepsilon \neq 0$. We obtain $\Theta_{1122}(T) = -3\varepsilon^2\lambda_1^4$; this term is sensitive to the Jordan normal form. Since $\varepsilon \neq 0$, $\lambda_1 = 0$ and $T(P)$ is nilpotent.

We now turn to the proof of Assertion 3. Suppose $\mathfrak{B} = 0$. Examining $\mathfrak{B}_{11}$ shows that $\Gamma_{11}{}^2 = 0$. Examining $\mathfrak{B}_{22}$ yields the remaining relation of Assertion 3-b. A direct computation shows that if the relations of Assertion 3-b are satisfied, then the Riemannian extension is Bach flat. $\qquad\square$

A special case of Theorem 15.4 occurs if the tensor field T is parallel. Indeed, a nilpotent $(1, 1)$-tensor field $T = \partial_{x^1} \otimes dx^2$ is parallel if and only if

$$\Gamma_{11}{}^1 = 0, \quad \Gamma_{11}{}^2 = 0, \quad \Gamma_{12}{}^1 = \Gamma_{22}{}^2, \quad \Gamma_{12}{}^2 = 0.$$

We apply Theorem 15.4 to derive the following result.

Corollary 15.5 *Let $\mathcal{M} = (M, \nabla)$ be an affine surface, let T be a parallel tensor field of Type $(1, 1)$, and let Φ be a symmetric $(0, 2)$-tensor field. Then $\mathcal{N} := (T^*M, g_{\nabla, \Phi, T})$ is Bach flat if and only if either $T = \mathrm{Id}$ or T is nilpotent.*

We have assumed in Theorem 15.4 either that T is a scalar multiple of the identity or that T is non-zero and not nilpotent. There are, however, examples where one can pass from one setting to the other. Let $M = \mathbb{R}^2$, let $\alpha(x^2)$ be a smooth real-valued function which vanishes to infinite-order at $x^2 = 0$ and which is positive for $x^2 \neq 0$. Assume that

$$\Gamma_{11}{}^2 = 0 \text{ and } (\Gamma_{11}{}^1)^2 - \Gamma_{11}{}^1 \Gamma_{12}{}^2 + \partial_{x^1}(\Gamma_{11}{}^1 - \Gamma_{12}{}^2) = 0.$$

Let

$$T(x^1, x^2) = \left\{ \begin{array}{ll} \begin{pmatrix} \alpha(x^2) & 0 \\ 0 & \alpha(x^2) \end{pmatrix} & \text{if } x^2 \leq 0 \\[3ex] \begin{pmatrix} 0 & \alpha(x^2) \\ 0 & 0 \end{pmatrix} & \text{if } x^2 \geq 0. \end{array} \right\}.$$

One may then compute that $\mathfrak{B} = 0$ so this yields a Bach flat manifold where the Jordan normal form of T changes at $x^2 = 0$. Furthermore, if we only assume that α is C^k for $k \geq 2$, we still obtain a solution; thus there is no hypo-ellipticity present when considering the solutions to the equations $\mathfrak{B} = 0$ and Remark 12.15 fails in this context.

15.1.5 QUASI-EINSTEIN MANIFOLDS.

Let $\mathcal{N} = (N, g)$ be a pseudo-Riemannian manifold. If the associated Ricci tensor ρ_g is a constant multiple of the metric, then $\mathcal{N}$ is said to be an *Einstein manifold*. If there is a smooth function h so that $e^h g$ is Einstein, then $\mathcal{N}$ is said to be a *conformally Einstein manifold*. More generally, let μ_N be a real constant and let F be a smooth function on N. Let $\mathcal{H}_g$ be the Hessian defined by the Levi-Civita connection. We say that $\mathcal{N} := (N, g, F, \mu_N)$ is a *quasi-Einstein manifold* if the following relation is satisfied:

$$\mathcal{H}_g F + \rho_g - \mu_N dF \otimes dF = \lambda g \quad \text{for some} \quad \lambda \in \mathbb{R}. \tag{15.1.e}$$

We say that $\mathcal{N}$ is *isotropic* if $\|dF\| = 0$. If λ is allowed to be a smooth function, then $\mathcal{N}$ is said to be a *generalized quasi-Einstein manifold*. We make the following observations:

Remark 15.6

1. If F is constant, then Equation (15.1.e) is the Einstein equation; quasi-Einstein manifolds are a generalization of Einstein manifolds.

2. If $n \geq 3$ and if $\mu_N = -\frac{1}{n-2}$ is the *critical eigenvalue*, then $\mathcal{N}$ is a generalized quasi-Einstein manifold if and only if $(N, e^{-\frac{2}{n-2}F} g)$ is Einstein. Thus any conformally Einstein manifold is quasi-Einstein with $\mu_N = -\frac{1}{n-2}$. For this reason we will often suppose $\mu_N \neq -\frac{1}{n-2}$. We refer to Brinkmann [3] or Gover and Nurowski [36] for further details concerning conformally Einstein manifolds.

3. For $\mu = 1$, the change of variable $h = e^{-F}$ yields $\mathcal{H}_g h - h\rho = -h\lambda g$. Let $\lambda = -\frac{\Delta h}{h}$. We then have $\mathcal{H}_g f - h\rho = (\Delta h)g$. This is the defining equation of the so-called *static manifolds* that arise in the study of static space-times as discussed in Kobayashi [38] and in Kobayashi and Obata [40].

4. If $\mu = 0$ and if λ is constant, one obtains the gradient Ricci soliton equation. This yields self-similar solutions of the Ricci flow. See, for example, the discussion in Cao and Chen [15] or in Munteanu and Sesum [45].

Let $\mathcal{M}$ be an affine surface. There is a useful change of variables.

Lemma 15.7 *Let $\mathcal{M} = (M, \nabla)$ be an affine manifold. Set $f = e^{-\mu h}$.*

1. $\mathcal{H}f - \mu f \rho_s = -\mu e^{-\mu h(P)}\{\mathcal{H}h + \rho_s - \mu dh \otimes dh\}$.
2. *If $\mu \neq 0$, then $f \in E(\mu, \mathcal{M})$ if and only if $\mathcal{H}h + \rho_s = \mu dh \otimes dh$.*

Proof. Fix a point P of M. Since ∇ is torsion-free, we may apply Lemma 9.1 of Book III to choose local coordinates so that $\Gamma_{ij}{}^k(P) = 0$. We then have

$$\{\mathcal{H}f\}_{ij}(P) = \partial_{x^i} \partial_{x^j}\{e^{-\mu h}\}(P) = e^{-\mu h(P)}\left\{-\mu\partial_{x^i}\partial_{x^j}h + \mu^2\partial_{x^i}h\partial_{x^j}h\right\}(P)$$
$$= e^{-\mu h(P)}\left\{-\mu\mathcal{H}h + \mu^2 dh \otimes dh\right\}_{ij}(P),$$

$$\{\mathcal{H}f - \mu f\rho_s\}_{ij}(P) = -\mu e^{-\mu h(P)}\{\mathcal{H}h + \rho_s - \mu dh \otimes dh\}_{ij}(P).$$

Assertion 1 follows since P was arbitrary. Since μ was non-zero, Assertion 2 follows from Assertion 1. □

We can now use the metric $g_{\nabla,\Phi}$ of Equation (15.1.b) to construct quasi-Einstein manifolds; it is this construction which motivated the terminology *quasi-Einstein equation* in affine geometry in the first instance.

Theorem 15.8 *Let $\mathcal{M}$ be an affine surface, let $\mu \neq 0$, let $f = e^{-\mu h} \in E(\mu, \mathcal{M})$, and let Φ be an arbitrary symmetric $(0, 2)$-tensor field on M. Then $\mathcal{N} := (T^*M, g_{\nabla,\Phi}, 2\pi^*h, \frac{1}{2}\mu)$ is a self-dual isotropic quasi-Einstein Walker manifold of signature $(2, 2)$ with $\lambda = 0$.*

Proof. Since $f \in E(\mu, \mathcal{M})$, $\mathcal{H}f - \mu f \rho_s = 0$ so $\mathcal{H}h + \rho_s - \mu dh \otimes dh = 0$ by Lemma 15.7. We apply Lemma 15.3 to see that

$$\mathcal{H}_g 2\pi^* h + \rho_g - \tfrac{1}{2}\mu \pi^*(d2h \otimes d2h) = \pi^*\{2\mathcal{H}h + 2\rho_s - 2\mu dh \otimes dh\} = 0\,.$$

We conclude that $(2\pi^* h, \tfrac{1}{2}\mu, \lambda = 0)$ satisfies Equation (15.1.e). $\square$

We note the factor of $\tfrac{1}{2}$ in passing from the eigenvalue μ on the base to the eigenvalue $\mu_N = \tfrac{1}{2}\mu$ on the cotangent bundle. The critical eigenvalue is $\mu = -\frac{1}{m-1}$ for the affine quasi-Einstein equation on an affine manifold as noted in Section 12.3. The corresponding critical eigenvalue is $\mu_N = -\tfrac{1}{2}\frac{1}{m-1} = -\frac{1}{n-2}$ on the cotangent space; this eigenvalue corresponds to the case $\mathcal{N}$ is conformally Einstein, as noted in Remark 15.6.

Theorem 15.8 gave an algorithm for constructing quasi-Einstein metrics on the cotangent bundle given a solution to the affine quasi-Einstein equation on the base. The following result of Brozos-Vázquez et al. [11] provides a partial converse to this result. We shall omit the proof as it is somewhat long and technical.

Theorem 15.9   Let $\mathcal{N} := (N, g, F, \tfrac{1}{2}\mu)$ be a self-dual isotropic quasi-Einstein manifold with λ constant, $\mu \neq -1$, and $\|\nabla F\|^2 = 0$. Fix a point P of N. There exists an affine surface $\mathcal{M}$, there exists $f = e^{-\mu h} \in E(\mu, \mathcal{M})$, and there exists a symmetric $(0, 2)$-tensor field on $\mathcal{M}$ so that near P, $\mathcal{N}$ is locally isomorphic to $(T^* M, g_{\nabla, \Phi}, 2\pi^* h, \tfrac{1}{2}\mu)$.

Based on Corollary 15.5 and proceeding in a similar way as in Theorem 15.8, one may construct Bach flat quasi-Einstein metrics. We use the nilpotent tensor T to define the tensor $\hat{\Phi}(X, Y) = \Phi(TX, TY)$. If $T = \partial_{x^1} \otimes dx^2$, then $\hat{\Phi} = \Phi_{11} dx^1 \otimes dx^1$, and one has:

Theorem 15.10   Let $\mathcal{M}$ be an affine surface and let T be a parallel nilpotent $(1, 1)$-tensor field on $\mathcal{M}$. Let $h \in C^\infty(M)$ be a smooth function. Then $\mathcal{N} := (T^* M, g_{\nabla, \Phi}, 2\pi^* h, \tfrac{1}{2}\mu)$ is an isotropic Bach flat quasi-Einstein manifold if and only if $dh(\ker\{T\}) = 0$ and the deformation tensor field Φ satisfies $\hat{\Phi} = -\{\mathcal{H}h + \rho_s - \mu dh \otimes dh\}$.

Theorem 15.10 can be further specialized so that the Bach flat structure is anti-self-dual. This shows that self-duality and anti-self-duality cannot be interchanged in Theorem 15.9. We refer to Brozos-Vázquez et al. [11] for the proof of the following result, which is an application of the distinguished coordinates considered in Section 12.2.

Corollary 15.11   Let $\mathcal{M}$ be an affine surface with symmetric and recurrent Ricci tensor of rank 1. Let T be a parallel nilpotent $(1, 1)$-tensor field on $\mathcal{M}$ and let $h \in C^\infty(M)$ be a smooth function with $dh(\ker\{T\}) = 0$ satisfying $\mathcal{H}h + \rho_s - \mu dh \otimes dh = 0$.

1. If the deformation tensor Φ vanishes, then $\mathcal{N} := (T^* M, g_{\nabla, \Phi}, 2\pi^* h, \tfrac{1}{2}\mu)$ is an isotropic Bach flat quasi-Einstein manifold.

2. *If the recurrence one-form ω given by $\nabla\rho = \omega \otimes \rho$ satisfies $\omega(\ker\{T\}) = 0$, then we have that $\mathcal{N} := (T^*M, g_{\nabla,\Phi}, 2\pi^*h, \frac{1}{2}\mu)$ is anti-self-dual for $\Phi = 0$.*

15.2 SIGNATURE $(2,2)$ VSI MANIFOLDS

Let $R_{i_1,\ldots,i_4;j_1,\ldots,j_k}$ be the components of the kth covariant derivative of the Levi-Civita connection of a pseudo-Riemannian manifold $\mathcal{N}$. Scalar invariants of the metric can be formed by using the metric tensors g^{ij} and g_{ij} to *fully contract* all indices. For example, the scalar curvature τ, the *norm of the Ricci tensor* $\|\rho\|^2$, and the *norm of the full curvature tensor* $\|R\|^2$ are scalar invariants which are given by:

$$\tau := g^{ij} R_{kij}{}^k, \qquad \|\rho\|^2 := g^{i_1 j_1} g^{i_2 j_2} R_{ki_1 i_2}{}^k R_{\ell j_1 j_2}{}^\ell, \quad \text{and}$$
$$\|R\|^2 := g^{i_1 j_1} g^{i_2 j_2} g^{i_3 j_3} g_{i_4 j_4} R_{i_1 i_2 i_3}{}^{i_4} R_{j_1 j_2 j_3}{}^{j_4}.$$

Such invariants are called *Weyl scalar invariants*; if all possible such invariants vanish, then $\mathcal{M}$ is said to be *VSI* (vanishing scalar invariants). We refer, for example, to Coley et al. [18] for further details. In the Riemannian setting, the vanishing of $\|R\|^2$ implies $\mathcal{N}$ is flat; this is not the case for pseudo-Riemannian manifolds as we shall see subsequently.

We take $X = 0$ (i.e., $a^i = 0$ for all i) and $S = T$ in Equation (15.1.c) to define

$$g_{\nabla,\Phi,T} := 2\,dx^i \circ dy_i + \left\{ \frac{1}{2} y_r y_s (T^r{}_i T^s{}_j + T^r{}_j T^s{}_i) - 2y_k \Gamma_{ij}{}^k + \Phi_{ij} \right\} dx^i \circ dx^j \qquad (15.2.a)$$

where $\xi \circ \eta := \frac{1}{2}(\xi \otimes \eta + \eta \otimes \xi)$ denotes the symmetric tensor product. In Section 15.2.1, we show that if $T = 0$ or if $0 \neq T$ and T is nilpotent, then $g_{\nabla,\Phi,T}$ is VSI. In Section 15.2.2, we construct scalar invariants of such manifolds which are not of Weyl type. In Section 15.2.3, we consider various examples defined by Type $\mathcal{A}$ or Type $\mathcal{B}$ models.

15.2.1 T IS NILPOTENT. If the underlying manifold is flat, then the Riemannian extension takes the form $g_{\nabla,\Phi} = 2dx^i \otimes dy_i + \Phi_{ij}(\vec{x})dx^i \otimes dx^j$. This is called a *plane wave manifold*; such manifolds are VSI by results of Gilkey and Nikčević [33]. We will show in Theorem 15.14 that suitable generalizations of this structure are VSI; these structures will play an important role in our discussion of Bach flat manifolds subsequently in Section 15.3. If T is nilpotent, we apply Lemma 15.2 to choose coordinates so the only non-zero component of T is $T^1{}_2 = 1$. To allow for the possibility that T could vanish identically, we shall suppose $T^1{}_2$ is constant in the following result. To simplify the notation, we let $x^3 = y_1$ and $x^4 = y_2$ henceforth.

Lemma 15.12 *Let $g = g_{\nabla,\Phi,T}$ where $T = c\,\partial_{x^1} \otimes dx^2$ where c is a real constant. Then the variables $\{g_{ij}, g^{ij}, {}^g\Gamma_{ij}{}^k, R_{abcd;e_1\ldots e_k}\}$ depend polynomially on the variables $\{\Gamma_{ij}{}^k, \Phi_{ij}, y_i\}$ and the derivatives of the variables $\{\Gamma_{ij}{}^k, \Phi_{ij}\}$ with respect to x^1 and x^2.*

1. $g_{11} = a_{11}(\vec{x}, \vec{y})$, $g_{12} = a_{12}(\vec{x}, \vec{y})$, $g_{22} = a_{22}(\vec{x}, \vec{y})$, $g_{13} = g_{24} = 1$,
   $g_{14} = g_{23} = g_{33} = g_{34} = g_{44} = 0$.

2. $g^{11} = g^{12} = g^{22} = g^{14} = g^{23} = 0$, $g^{13} = g^{24} = 1$, $g^{33} = -a_{11}(\vec{x}, \vec{y})$,
   $g^{34} = -a_{12}(\vec{x}, \vec{y})$, $g^{44} = -a_{22}(\vec{x}, \vec{y})$.

3. *The only possibly non-zero Christoffel symbols of the Levi-Civita connection of g are* ${}^g\Gamma_{11}{}^1$,
   ${}^g\Gamma_{11}{}^2$, ${}^g\Gamma_{11}{}^3$, ${}^g\Gamma_{11}{}^4$, ${}^g\Gamma_{12}{}^1$, ${}^g\Gamma_{12}{}^2$, ${}^g\Gamma_{12}{}^3$, ${}^g\Gamma_{12}{}^4$, ${}^g\Gamma_{13}{}^3$, ${}^g\Gamma_{13}{}^4$, ${}^g\Gamma_{14}{}^3$, ${}^g\Gamma_{14}{}^4$, ${}^g\Gamma_{22}{}^1$,
   ${}^g\Gamma_{22}{}^2$, ${}^g\Gamma_{22}{}^3$, ${}^g\Gamma_{22}{}^4$, ${}^g\Gamma_{23}{}^3$, ${}^g\Gamma_{23}{}^4$, ${}^g\Gamma_{24}{}^3$, ${}^g\Gamma_{24}{}^4$.

4. *The only possible non-zero components of the curvature tensor are* R_{1212}, R_{1213},
   R_{1214}, R_{1223}, R_{1224}, *and* R_{2323}.

Proof. The first assertion is immediate from Equation (15.2.a). If we write, schematically,

$$g = \begin{pmatrix} a & I \\ I & 0 \end{pmatrix} \quad \text{then} \quad g^{-1} = \begin{pmatrix} 0 & I \\ I & -a \end{pmatrix},$$

then Assertion 2 follows. We use the Koszul formula (see Theorem 3.7 of Book I) to see that ${}^g\Gamma_{abc} = 0$ unless at least two indices belong to $\{1, 2\}$. Thus

$$0 = {}^g\Gamma_{133} = {}^g\Gamma_{331} = {}^g\Gamma_{134} = {}^g\Gamma_{143} = {}^g\Gamma_{341} = {}^g\Gamma_{144} = {}^g\Gamma_{441},$$
$$0 = {}^g\Gamma_{233} = {}^g\Gamma_{332} = {}^g\Gamma_{234} = {}^g\Gamma_{243} = {}^g\Gamma_{342} = {}^g\Gamma_{244} = {}^g\Gamma_{442},$$
$$0 = {}^g\Gamma_{333} = {}^g\Gamma_{334} = {}^g\Gamma_{343} = {}^g\Gamma_{444} = {}^g\Gamma_{344} = {}^g\Gamma_{443}.$$

We raise indices using Assertion 2 to see

$$0 = {}^g\Gamma_{13}{}^1 = {}^g\Gamma_{13}{}^2 = {}^g\Gamma_{14}{}^1 = {}^g\Gamma_{14}{}^2 = {}^g\Gamma_{24}{}^1 = {}^g\Gamma_{24}{}^2 = {}^g\Gamma_{23}{}^1,$$
$$0 = {}^g\Gamma_{23}{}^2 = {}^g\Gamma_{33}{}^1 = {}^g\Gamma_{33}{}^2 = {}^g\Gamma_{33}{}^3 = {}^g\Gamma_{33}{}^4 = {}^g\Gamma_{34}{}^1 = {}^g\Gamma_{34}{}^2,$$
$$0 = {}^g\Gamma_{34}{}^3 = {}^g\Gamma_{34}{}^4 = {}^g\Gamma_{44}{}^1 = {}^g\Gamma_{44}{}^2 = {}^g\Gamma_{44}{}^3 = {}^g\Gamma_{44}{}^4.$$

The Levi-Civita connection in dimension 4 has 40 components. We have listed 20; the remaining 20 appear in Assertion 3. A direct computation shows

$$0 = R_{1234} = R_{1313} = R_{1314} = R_{1323} = R_{1324} = R_{1334} = R_{1414},$$
$$0 = R_{1423} = R_{1424} = R_{1434} = R_{2324} = R_{2334} = R_{2424} = R_{2434} = R_{3434}.$$

The curvature tensor has 21 components in dimension 4. We have listed 15; the remaining 6 appear in Assertion 4. $\square$

Lemma 15.12 controls which curvature tensors are potentially non-zero. We introduce some additional notation to control the higher covariant derivatives of the curvature tensor. Let $\mathfrak{o}(\cdot)$ be the maximal-order of an expression in the dual variables $\{y_1 = x^3, y_2 = x^4\}$. Thus if $\mathfrak{o}(\cdot) = 0$, these variables do not occur, if $\mathfrak{o}(\cdot) = 1$, the expression is linear in the variables $\{x_3, x_4\}$, and so forth. If $\mathfrak{o}(R_{ijk\ell}) = 2$, then $R_{ijk\ell}$ is at most quadratic in $\{x_3, x_4\}$; if $\mathfrak{o}(R_{ijk\ell}) = 1$, then $R_{ijk\ell}$ is at most linear in $\{x_3, x_4\}$; and if $\mathfrak{o}(R_{ijk\ell}) = 0$, then $R_{ijk\ell}$ does not involve $\{x_3, x_4\}$. A direct computation shows

$$\mathfrak{o}({}^g\Gamma_{11}{}^1) = 0, \quad \mathfrak{o}({}^g\Gamma_{11}{}^2) = 0, \quad \mathfrak{o}({}^g\Gamma_{11}{}^3) = 1, \quad \mathfrak{o}({}^g\Gamma_{11}{}^4) = 2,$$
$$\mathfrak{o}({}^g\Gamma_{12}{}^1) = 0, \quad \mathfrak{o}({}^g\Gamma_{12}{}^2) = 0, \quad \mathfrak{o}({}^g\Gamma_{12}{}^3) = 1, \quad \mathfrak{o}({}^g\Gamma_{12}{}^4) = 2,$$
$$\mathfrak{o}({}^g\Gamma_{13}{}^3) = 0, \quad \mathfrak{o}({}^g\Gamma_{13}{}^4) = 0, \quad \mathfrak{o}({}^g\Gamma_{14}{}^3) = 0, \quad \mathfrak{o}({}^g\Gamma_{14}{}^4) = 0,$$
$$\mathfrak{o}({}^g\Gamma_{22}{}^1) = 1, \quad \mathfrak{o}({}^g\Gamma_{22}{}^2) = 0, \quad \mathfrak{o}({}^g\Gamma_{22}{}^3) = 2, \quad \mathfrak{o}({}^g\Gamma_{22}{}^4) = 2,$$
$$\mathfrak{o}({}^g\Gamma_{23}{}^3) = 0, \quad \mathfrak{o}({}^g\Gamma_{23}{}^4) = 1, \quad \mathfrak{o}({}^g\Gamma_{24}{}^3) = 0, \quad \mathfrak{o}({}^g\Gamma_{24}{}^4) = 0,$$
$$\mathfrak{o}(R_{1212}) = 2, \quad \mathfrak{o}(R_{1213}) = 1, \quad \mathfrak{o}(R_{1214}) = 0, \quad \mathfrak{o}(R_{1223}) = 1,$$
$$\mathfrak{o}(R_{1224}) = 1, \quad \mathfrak{o}(R_{2323}) = 0.$$

We define the *defect* by setting

$$\mathfrak{d}({}^g\Gamma_{ij}{}^k) = -\sum_{n=1}^{2}\{\delta_{i,n} + \delta_{j,n} - \delta_{k,n}\} + \sum_{n=3}^{4}\{\delta_{i,n} + \delta_{j,n} - \delta_{k,n}\},$$

$$\mathfrak{d}(R_{i_1 i_2 i_3 i_4; i_5 \dots i_\nu}) := \sum_{n=1}^{\nu}\{\delta_{i_n,3} + \delta_{i_n,4} - \delta_{i_n,1} - \delta_{i_n,2}\}.$$

In brief, we count, with multiplicity, each lower index "1" or "2" with the sign -1 and "3" or "4" with the sign $+1$ and, dually, reverse the sign for upper indices. Set $\mathfrak{x} = \mathfrak{o} + \mathfrak{d}$ and compute:

$$\mathfrak{x}({}^g\Gamma_{11}{}^1) = -1, \quad \mathfrak{x}({}^g\Gamma_{11}{}^2) = -1, \quad \mathfrak{x}({}^g\Gamma_{11}{}^3) = -2, \quad \mathfrak{x}({}^g\Gamma_{11}{}^4) = -1,$$
$$\mathfrak{x}({}^g\Gamma_{12}{}^1) = -1, \quad \mathfrak{x}({}^g\Gamma_{12}{}^2) = -1, \quad \mathfrak{x}({}^g\Gamma_{12}{}^3) = -2, \quad \mathfrak{x}({}^g\Gamma_{12}{}^4) = -1,$$
$$\mathfrak{x}({}^g\Gamma_{13}{}^3) = -1, \quad \mathfrak{x}({}^g\Gamma_{13}{}^4) = -1, \quad \mathfrak{x}({}^g\Gamma_{14}{}^3) = -1, \quad \mathfrak{x}({}^g\Gamma_{14}{}^4) = -1,$$
$$\mathfrak{x}({}^g\Gamma_{22}{}^1) = 0, \quad \ \ \mathfrak{x}({}^g\Gamma_{22}{}^2) = -1, \quad \mathfrak{x}({}^g\Gamma_{22}{}^3) = -1, \quad \mathfrak{x}({}^g\Gamma_{22}{}^4) = -1, \qquad (15.2.\text{b})$$
$$\mathfrak{x}({}^g\Gamma_{23}{}^3) = -1, \quad \mathfrak{x}({}^g\Gamma_{23}{}^4) = 0, \quad \ \ \mathfrak{x}({}^g\Gamma_{24}{}^3) = -1, \quad \mathfrak{x}({}^g\Gamma_{24}{}^4) = -1,$$
$$\mathfrak{x}(R_{1212}) = -2, \quad \mathfrak{x}(R_{1213}) = -1, \quad \mathfrak{x}(R_{1214}) = -2, \quad \mathfrak{x}(R_{1223}) = -1,$$
$$\mathfrak{x}(R_{1224}) = -1, \quad \mathfrak{x}(R_{2323}) = 0.$$

We can now control the non-zero higher covariant derivatives.

Lemma 15.13   *Suppose that* $R_{i_1 i_2 i_3 i_4; i_5 \dots i_\nu} \neq 0$. *Then* $\mathfrak{x}(R_{i_1 i_2 i_3 i_4; i_5 \dots i_\nu}) \leq 0$. *Furthermore,* $\mathfrak{x}(R_{i_1 i_2 i_3 i_4; i_5 \dots i_\nu}) = 0$ *if and only if* $R_{i_1 i_2 i_3 i_4; i_5 \dots i_\nu} = \pm R_{2323}$.

Proof.   Let $R_{ijk\ell} \neq 0$. By Equation (15.2.b), we have $\mathfrak{x}(R_{ijk\ell}) \leq 0$ with equality if and only if $R_{ijk\ell} = \pm R_{2323}$. This establishes the result if $\nu = 4$. Next we suppose $\nu = 5$ and examine ${}^g\nabla R$. By Lemma 15.12, $R_{i_1 i_2 i_3 i_4; n}$ is polynomial in its arguments; we suppose this polynomial does not vanish identically and that $\mathfrak{r}(R_{i_1 i_2 i_3 i_4;n}) \geq 0$. We argue for a contradiction. Expand

$$R_{i_1 i_2 i_3 i_4;n} = \partial_n R_{i_1 i_2 i_3 i_4} - \sum_a {}^g\Gamma_{ni_1}{}^a R_{a i_2 i_3 i_4} - \sum_a {}^g\Gamma_{ni_2}{}^a R_{i_1 a i_3 i_4}$$
$$- \sum_a {}^g\Gamma_{ni_3}{}^a R_{i_1 i_2 a i_4} - \sum_a {}^g\Gamma_{ni_4}{}^a R_{i_1 i_2 i_3 a}.$$

Since $R_{i_1 i_2 i_3 i_4;n}$ does not vanish identically, at least one of the terms in the previous display is non-zero. We distinguish cases.

Case 1. Suppose that $\partial_n R_{i_1 i_2 i_3 i_4} \neq 0$. If $n \in \{1, 2\}$, then

$$\mathfrak{d}(R_{i_1 i_2 i_3 i_4;n}) = \mathfrak{d}(R_{i_1 i_2 i_3 i_4}) - 1 < 0, \quad \mathfrak{o}(\partial_n R_{i_1 i_2 i_3 i_4}) \leq \mathfrak{o}(R_{i_1 i_2 i_3 i_4}),$$
$$\mathfrak{r}(\partial_n R_{i_1 i_2 i_3 i_4}) \leq \mathfrak{r}(R_{i_1 i_2 i_3 i_4}) - 1 < 0.$$

If $n \in \{3, 4\}$, then necessarily $\mathfrak{o}(R_{i_1 i_2 i_3 i_4}) > 0$ to ensure $R_{i_1 i_2 i_3 i_4}$ in fact depends on (x_3, x_4). Thus $R_{i_1 i_2 i_3 i_4} \neq R_{2323}$. We have

$$\mathfrak{d}(R_{i_1 i_2 i_3 i_4;n}) = \mathfrak{d}(R_{i_1 i_2 i_3 i_4}) + 1, \quad \mathfrak{o}(\partial_n R_{i_1 i_2 i_3 i_4}) \leq \mathfrak{o}(R_{i_1 i_2 i_3 i_4}) - 1,$$
$$\mathfrak{r}(R_{i_1 i_2 i_3 i_4;n}) \leq \mathfrak{r}(R_{i_1 i_2 i_3 i_4}) < 0.$$

Thus in any event, $\mathfrak{r}(R_{i_1 i_2 i_3 i_4;n}) < 0$, which is false.

Case 2. Suppose that ${}^g\Gamma_{ni_1}{}^a R_{a i_2 i_3 i_4} \geq 0$ for some a; the remaining four cases involving ${}^g\Gamma_{ni_2}{}^a R_{i_1 a i_3 i_4}$, ${}^g\Gamma_{ni_3}{}^a R_{i_1 i_2 a i_4}$, and ${}^g\Gamma_{ni_4}{}^a R_{i_1 i_2 i_3 a}$ are similar. Since $0 \geq \mathfrak{r}({}^g\Gamma_{ni_1}{}^a)$ and since $0 \geq \mathfrak{r}(R_{a i_2 i_3 i_4})$, we have that

$$0 \geq \mathfrak{r}({}^g\Gamma_{ni_1}{}^a) + \mathfrak{r}(R_{a i_2 i_3 i_4}) = \mathfrak{r}({}^g\Gamma_{ni_1}{}^a R_{a i_2 i_3 i_4}) \geq 0.$$

Thus $\mathfrak{r}({}^g\Gamma_{ni_1}{}^a) = 0$ and $\mathfrak{r}(R_{a i_2 i_3 i_4}) = 0$. By Equation (15.2.b), $R_{a i_2 i_3 i_4} = \pm R_{2323}$ so $a = 2$ or $a = 3$. Since $\mathfrak{r}({}^g\Gamma_{ni_1}{}^a) = 0$, Equation (15.2.b) then shows ${}^g\Gamma_{ni_1}{}^a$ is either ${}^g\Gamma_{22}{}^1$ or ${}^g\Gamma_{23}{}^4$. This shows that $a = 1$ or $a = 4$. This provides the desired contradiction and establishes the desired result for the components of ${}^g\nabla R$. The argument for ${}^g\nabla^\ell R$ for $\ell \geq 2$ now proceeds by induction; we do not have the additional complexity involved in considering variables $R_{i_1 i_2 i_3 i_4;i_5 \ldots i_\nu}$ where $\mathfrak{r}(R_{i_1 i_2 i_3 i_4;i_5 \ldots i_\nu}) = 0$. □

We can extend the results of Gilkey and Nikčević [33] from the setting of generalized plane wave manifolds to the setting at hand as follows.

Theorem 15.14   *Let $\mathcal{N} = (T^*M, g_{\nabla,\Phi,T})$. Suppose either T vanishes identically or T never vanishes. The following assertions are equivalent:*

(1) T is nilpotent.   (2) $\mathcal{N}$ is VSI.   (3) $\|R\|^2 = \|\rho\|^2 = 0$.   (4) $\|\rho\|^2 = \tau = 0$.

Proof. Suppose T is nilpotent. Let $\mathcal{W}$ be a Weyl scalar invariant formed from the curvature tensor and its covariant derivatives. By Lemma 15.12, we can contract an index "1" against an index "3" and an index "2" against an index "4". We can also contract indices $\{3,4\}$ against $\{3,4\}$. We may not, however, contract indices $\{1,2\}$ against indices $\{1,2\}$. Consequently, if $A = R_{i_1 i_2 i_3 i_4 ; i_5 \ldots i_\nu} \cdots$ is a monomial, then $\deg_1(A) \le \deg_3(A)$ and $\deg_2(A) \le \deg_4(A)$. The inequality can, of course, be strict as we can also contract an index 3 or 4 against an index 3 or 4. This implies that $\mathfrak{d}(A) \ge 0$. Since $\mathfrak{o}(A) \ge 0$, this implies $\mathfrak{r}(A) \ge 0$. By Lemma 15.13, $\mathfrak{r}(A) \le 0$. Thus we have $\mathfrak{r}(A) = 0$. This implies A is a power of R_{2323}. We cannot contract an index "2" against an index "3". Consequently, $\mathcal{W} = 0$. Thus Assertion 1 implies Assertion 2. It is immediate that Assertion 2 implies Assertion 3 and Assertion 4 since if $\mathcal{N}$ is VSI, we have $\|R\|^2 = \|\rho\|^2 = \tau = 0$.

Fix P and $\{\lambda_1, \lambda_2\}$ be the (possibly complex) eigenvalues of $T(P)$. As in the proof of Theorem 15.4, we can make a complex linear change of coordinates in the $\{x^1, x^2\}$ variables to put T in Jordan normal form and express

$$T(P) := \begin{pmatrix} \lambda_1 & \varepsilon \\ 0 & \lambda_2 \end{pmatrix}.$$

This induces a corresponding dual complex linear change of coordinates in the $\{y_1, y_2\}$ variables. The parameter ε plays no role and we obtain at P that

$$\tau = 2\left(\lambda_1^2 + \lambda_1 \lambda_2 + \lambda_2^2\right), \qquad \|R\|^2 = 4(\lambda_1^4 + \lambda_1^2 \lambda_2^2 + \lambda_2^4),$$
$$\|\rho\|^2 = 2\lambda_1^4 + 2\lambda_1^3 \lambda_2 + \lambda_1^2 \lambda_2^2 + 2\lambda_1 \lambda_2^3 + 2\lambda_2^4.$$

Suppose that $\|R\|^2 = \|\rho\|^2 = 0$. If the eigenvalues are real, then the vanishing of $\|R\|^2$ implies $(\lambda_1, \lambda_2) = (0,0)$ so T is nilpotent. Thus we assume the eigenvalues are complex. Consequently, $\lambda_2 = \bar{\lambda}_1 \ne 0$. Set $\lambda_1 = re^{i\theta}$ and $\lambda_2 = re^{-i\theta}$ for $r \ne 0$. The equations in question are homogeneous so we may assume without loss of generality $r = 1$. We have

$$0 = \|\rho\|^2 - \tfrac{1}{2}\|R\|^2 = 2\lambda_1^3 \lambda_2 - \lambda_1^2 \lambda_2^2 + 2\lambda_1 \lambda_2^3.$$

Dividing this equation by $\lambda_1 \lambda_2$ yields $0 = 2\lambda_1^2 - \lambda_1 \lambda_2 + 2\lambda_2^2$. Setting $\lambda_1 = e^{i\theta}$ and $\lambda_2 = e^{-i\theta}$, we have $0 = e^{4i\theta} + 1 + e^{-4i\theta}$ so $\cos(4\theta) = -\tfrac{1}{2}$ and $0 = 2e^{2i\theta} - 1 + 2e^{2i\theta}$. Thus $\cos(2\theta) = \tfrac{1}{4}$. The angle addition formulas yield $-\tfrac{1}{2} = \cos(4\theta) = 2\cos^2(2\theta) - 1 = \tfrac{1}{8} - 1$ so this case does not occur. Thus Assertion 3 implies Assertion 1.

Assume $\tau = 0$ and $\|\rho\|^2 = 0$. We compute

$$\|\rho\|^2 - \tfrac{1}{2}\tau^2 = -\lambda_1 \lambda_2 (2\lambda_1 + \lambda_2)(\lambda_1 + 2\lambda_2).$$

Setting $\tau = 0$ then yields $\lambda_1 = \lambda_2 = 0$. Thus T is nilpotent and Assertion 4 implies Assertion 1. $\qquad\square$

The conditions of Assertion 3 and Assertion 4 are optimal in a certain sense. For θ constant and $r(x^1, x^2) \neq 0$, define an endomorphism of the tangent bundle which is not nilpotent by setting

$$T(x) := r(x^1, x^2) \begin{pmatrix} \cos(\theta) & \sin(\theta) \\ -\sin(\theta) & \cos(\theta) \end{pmatrix}.$$

Then $\tau = 2r^2(2\cos(2\theta) + 1)$ and $\|R\|^2 = 4r^4(2\cos(4\theta) + 1)$. Taking $\theta = \frac{\pi}{3}$ then yields $\|R\|^2 = \tau = 0$. Similarly,

$$\|\rho\|^2 = r^4(4\cos(4\theta) + 4\cos(2\theta) + 1) = 0 \text{ for } \theta = \frac{1}{2}\arctan\left(\frac{\sqrt{7}+1}{\sqrt{7}-1}\right).$$

Thus the conditions $\{\|R\|^2 = 0, \tau = 0\}$ or $\|\rho\|^2 = 0$ do not suffice to show T is nilpotent.

15.2.2 INVARIANTS OF VSI MANIFOLDS NOT OF WEYL TYPE.

We continue the discussion of Section 15.2.1. Let $\mathcal{M} = (M, \nabla)$ be an affine surface, let $0 \neq T$ be nilpotent, and let $\mathcal{N} := (T^*M, g_{\nabla,\Phi,T})$ be the associated VSI Riemannian extension. Apply Lemma 15.2 to choose local coordinates so that $T = \partial_{x^1} \otimes dx^2$. Take dual coordinates $y_1 = x^3$ and $y_2 = x^4$. Let $\{R, \rho\}$ denote the curvature operator and Ricci tensor of $\mathcal{N}$ and let $\{R^\nabla, \rho^\nabla, \rho_a^\nabla, \rho_s^\nabla\}$ be the curvature operator, Ricci tensor, alternating Ricci tensor, and symmetric Ricci tensor of $\mathcal{M}$, respectively. Let $\mathcal{V} := \text{span}\{\partial_{x^3}, \partial_{x^4}\}$ be the "vertical" space and let $\mathfrak{H} := \text{span}\{\partial_{x^1}, \partial_{x^2}\}$ be the "horizontal" space. The vertical space $\mathcal{V}$ is the kernel of $\pi_* : T_*(TN) \to TM$; this is the Walker distribution and is a maximal parallel null 2-dimensional subspace. The horizontal space $\mathcal{H}$, however, is not invariantly defined. We may then decompose

$$R(X, Y) = \begin{pmatrix} R^{\mathfrak{H}}_{\mathfrak{H}} = \begin{pmatrix} R_{XY1}{}^1 & R_{XY2}{}^1 \\ R_{XY1}{}^2 & R_{XY2}{}^2 \end{pmatrix} & R^{\mathfrak{H}}_{\mathcal{V}} = \begin{pmatrix} R_{XY3}{}^1 & R_{XY4}{}^1 \\ R_{XY3}{}^2 & R_{XY4}{}^2 \end{pmatrix} \\ R^{\mathcal{V}}_{\mathfrak{H}} = \begin{pmatrix} R_{XY1}{}^3 & R_{XY2}{}^3 \\ R_{XY1}{}^4 & R_{XY2}{}^4 \end{pmatrix} & R^{\mathcal{V}}_{\mathcal{V}} = \begin{pmatrix} R_{XY3}{}^3 & R_{XY4}{}^3 \\ R_{XY3}{}^4 & R_{XY4}{}^4 \end{pmatrix} \end{pmatrix}.$$

The following result follows by a direct computation.

Lemma 15.15   Let $\mathcal{N} := (T^*M, g_{\nabla,\Phi,T})$ where $T = \partial_{x^1} \otimes dx^2$.

1. $R^{\mathfrak{H}}_{\mathcal{V}}(X, Y) = 0$ for all (X, Y), i.e., $R_{abi}{}^j = 0$ for $3 \leq i \leq 4$, $1 \leq j \leq 2$.
2. $\{R^{\mathfrak{H}}_{\mathfrak{H}} + (R^{\mathcal{V}}_{\mathcal{V}})^t\}(X, Y) = 0$ for all (X, Y) i.e., $R_{ab1}{}^1 + R_{ab3}{}^3 = 0$, $R_{ab2}{}^2 + R_{ab4}{}^4 = 0$, $R_{ab1}{}^2 + R_{ab3}{}^4 = 0$, and $R_{ab2}{}^1 + R_{ab3}{}^4 = 0$.
3. $R^{\mathfrak{H}}_{\mathfrak{H}}(\partial_{x^i}, \partial_{x^j}) = 0$ for $i < j$ and $(i, j) \notin \{(1, 2), (2, 3)\}$.
4. $\begin{pmatrix} R_{231}{}^1 & R_{232}{}^1 \\ R_{231}{}^2 & R_{232}{}^2 \end{pmatrix} = \begin{pmatrix} 0 & 1 \\ 0 & 0 \end{pmatrix}.$
5. $\begin{pmatrix} R_{121}{}^1 & R_{122}{}^1 \\ R_{121}{}^2 & R_{122}{}^2 \end{pmatrix} = \begin{pmatrix} R^{\nabla}_{121}{}^1 & R^{\nabla}_{122}{}^1 \\ R^{\nabla}_{121}{}^2 & R^{\nabla}_{122}{}^2 \end{pmatrix} - x_3 \begin{pmatrix} -\Gamma_{11}{}^2 & \Gamma_{11}{}^1 - \Gamma_{12}{}^2 \\ 0 & \Gamma_{11}{}^2 \end{pmatrix}.$

6. $\mathrm{Tr}\{R_{\mathfrak{H}}^{\mathfrak{H}}(X,Y)\} = -2(\pi^{*}\rho_{a}^{\nabla})(X,Y)$.

7. $\rho_{ij} = 0$ if $i \geq 3$ or $j \geq 3$.

8. $\begin{pmatrix} \rho_{11} & \rho_{21} \\ \rho_{12} & \rho_{22} \end{pmatrix} = 2\rho_{s}^{\nabla} + \begin{pmatrix} 0 & 2x_{3}\Gamma_{11}{}^{2} \\ 2x_{3}\Gamma_{11}{}^{2} & -4x_{3}\Gamma_{11}{}^{1} - 2x_{4}\Gamma_{11}{}^{2} + 2x_{3}\Gamma_{12}{}^{2} + \Phi_{11} \end{pmatrix}$.

9. $^{g}\nabla R(i,j,1,1;k) + {}^{g}\nabla R(i,j,2,2;k) = 0$ unless $\{i,j,k\} \in \{1,2\}$.

Generically, $\mathcal{V}$ is the only parallel null distribution. Consequently, it defines an additional piece of structure. The distribution $\mathrm{span}\{\partial_{x^{1}}, \partial_{x^{2}}\}$ is not invariantly defined. To obtain an invariant object, we set $\tilde{\mathfrak{H}} := TN/\mathcal{V}$ and let $\tilde{\pi} : TN \to \tilde{\mathfrak{H}}$ be the natural projection. By Lemma 15.15, $\tilde{\pi}R(X,Y)v = 0$ for $v \in \mathcal{V}$ and thus $\tilde{\pi}R(X,Y)$ descends to a well defined map $R_{\tilde{\mathfrak{H}}}^{\tilde{\mathfrak{H}}}(X,Y)$ of $\tilde{\mathfrak{H}}$. Let $\{X_{3}, X_{4}\}$ be a local frame for $\mathcal{V}$. Choose $\{X_{1}, X_{2}\}$ so that

$$g(X_{1}, X_{3}) = g(X_{2}, X_{4}) = 1 \quad \text{and} \quad g(X_{1}, X_{4}) = g(X_{2}, X_{3}) = 0. \tag{15.2.c}$$

We note that $\{X_{1}, X_{2}\}$ is not uniquely defined by these relations as we can add an arbitrary element of $\mathcal{V}$ to either X_{1} or X_{2} and preserve Equation (15.2.c). However, $\{\tilde{\pi}X_{1}, \tilde{\pi}X_{2}\}$ is uniquely defined by Equation (15.2.c). And, in particular, if we take $X_{3} = \partial_{x_{3}}$ and $X_{4} = \partial_{x_{4}}$, then we may take $X_{1} = \partial_{x^{1}}$ and $X_{2} = \partial_{x^{2}}$.

Definition 15.16   We use Lemma 15.15 to introduce some additional quantities.

1. Since $\rho(X,Y) = 0$ if either X or Y belongs to $\mathcal{V}$, ρ descends to a map from $\tilde{\mathfrak{H}} \oplus \tilde{\mathfrak{H}}$ to $\mathbb{R}$ that we shall denote by $\rho^{\tilde{\mathfrak{H}}} \in S^{2}(\tilde{\mathfrak{H}}^{*})$. Let $\tilde{\pi} : T^{*}M \to M$. Because $\tilde{\pi}_{*}(\mathcal{V}) = 0$, $\tilde{\pi}_{*}$ induces a map from $\tilde{\mathfrak{H}}$ to TM. If $\Gamma_{11}{}^{2} = 0$, if $2\Gamma_{11}{}^{2} = \Gamma_{12}{}^{2}$, and if $\Phi_{11} = 0$, then $\rho^{\tilde{\mathfrak{H}}} = 2\tilde{\pi}^{*}\rho_{s}^{\nabla}$.

2. Let $\Omega(X,Y) = \mathrm{Tr}\{R_{\tilde{\mathfrak{H}}}^{\tilde{\mathfrak{H}}}(X,Y)\}$. Then $\Omega(X,Y) = 0$ if either X or Y belongs to $\mathcal{V}$ so Ω descends to an alternating bilinear map from $\tilde{\mathfrak{H}} \oplus \tilde{\mathfrak{H}}$ to $\mathbb{R}$ that we denote by $\Omega^{\tilde{\mathfrak{H}}} \in \Lambda^{2}(\tilde{\mathfrak{H}}^{*})$. We have $\Omega^{\tilde{\mathfrak{H}}} = -2\tilde{\pi}^{*}\rho_{a}^{\nabla}$.

3. As $\mathcal{V}$ is parallel, $^{g}\nabla R(X,Y;Z)$ maps $\mathcal{V}$ to $\mathcal{V}$. Consequently, $^{g}\nabla R(X,Y;Z)$ extends to an endomorphism $(^{g}\nabla R)^{\tilde{\mathfrak{H}}}(X,Y;Z)$ of $\tilde{\mathfrak{H}}$. A direct computation shows that $\mathrm{Tr}\{(^{g}\nabla R)^{\tilde{\mathfrak{H}}}(X,Y;Z)\} = 0$ if X, Y, or Z belongs to $\mathcal{V}$. We may therefore regard $\mathrm{Tr}\{(^{g}\nabla R)^{\tilde{\mathfrak{H}}}(X,Y;Z)\} \in \Lambda^{2}(\tilde{\mathfrak{H}}) \otimes \tilde{\mathfrak{H}}^{*}$. Assuming that $\Omega^{\tilde{\mathfrak{H}}} \neq 0$, we may decompose $\mathrm{Tr}\{(^{g}\nabla R)^{\tilde{\mathfrak{H}}}\} = \omega^{\tilde{\mathfrak{H}}} \otimes \Omega^{\tilde{\mathfrak{H}}}$ for $\omega^{\tilde{\mathfrak{H}}} \in \tilde{\mathfrak{H}}^{*}$. Moreover, one has $d\omega^{\tilde{\mathfrak{H}}} = \Omega^{\tilde{\mathfrak{H}}}$.

4. Suppose that we are at a point of $\mathcal{N}$ where $\rho^{\tilde{\mathfrak{H}}}$ defines a non-degenerate symmetric bilinear form on $\tilde{\mathfrak{H}}$. We may then define

$$\beta_{1} := \|\Omega^{\tilde{\mathfrak{H}}}\|_{\rho^{\tilde{\mathfrak{H}}}}^{2} = \frac{(R_{121}{}^{1} + R_{122}{}^{2})^{2}}{\rho_{11}\rho_{22} - \rho_{12}\rho_{12}}.$$

If we also assume that $\Omega^{\tilde{\mathfrak{H}}} \neq 0$ (i.e., $\rho_{a}^{\nabla} \neq 0$) or, equivalently, that $\beta_{1} \neq 0$, then $\omega^{\tilde{\mathfrak{H}}}$ is well defined and we may set $\beta_{2} := \|\omega^{\tilde{\mathfrak{H}}}\|_{\rho^{\tilde{\mathfrak{H}}}}^{2}$. We have that $\omega_{1}^{\tilde{\mathfrak{H}}} = \frac{R_{121}{}^{1};_{1} + R_{122}{}^{2};_{1}}{R_{121}{}^{1} + R_{122}{}^{2}}$ and that

$$\omega_2^{\tilde{\mathfrak{H}}} = \frac{R_{121}{}^1{}_{;2} + R_{122}{}^2{}_{;2}}{R_{121}{}^1 + R_{122}{}^2}. \text{ Consequently,}$$

$$\beta_2 := \frac{\rho_{22}^{\tilde{\mathfrak{H}}} \omega_1^{\tilde{\mathfrak{H}}} \omega_1^{\tilde{\mathfrak{H}}} + \rho_{11}^{\tilde{\mathfrak{H}}} \omega_2^{\tilde{\mathfrak{H}}} \omega_2^{\tilde{\mathfrak{H}}} - 2\rho_{12}^{\tilde{\mathfrak{H}}} \omega_1^{\tilde{\mathfrak{H}}} \omega_2^{\tilde{\mathfrak{H}}}}{\rho_{11}^{\tilde{\mathfrak{H}}} \rho_{22}^{\tilde{\mathfrak{H}}} - \rho_{12}^{\tilde{\mathfrak{H}}} \rho_{12}^{\tilde{\mathfrak{H}}}}.$$

15.2.3 BACH FLAT VSI MANIFOLDS.

It is obvious from the discussion given in Section 15.2.2 that β_1 and β_2 are isometry invariants of $\mathcal{N}$ where defined. Generically, β_1 and β_2 are very complicated expressions which involve non-trivial dependence on the fiber variables and which involve the endomorphism Φ. These invariants are uninteresting if $\Omega^{\tilde{\mathfrak{H}}} = 0$ or, equivalently in the setting where $g = g_{\nabla,\Phi,T}$ if ρ^∇ is symmetric because $\Omega^{\tilde{\mathfrak{H}}} = -2\pi^*\rho_a^\nabla$. Thus the Type $\mathcal{A}$ geometries do not lead to interesting examples in this setting. Let $\Gamma_{ij}{}^k = \frac{1}{x^1} A_{ij}{}^k$ define a Type $\mathcal{B}$ geometry and let $\mathcal{N} := (T^*M, g_{\nabla,\Phi,T})$. By Equation (14.1.a), the Ricci tensor is symmetric if and only if $A_{12}{}^1 + A_{22}{}^2 = 0$. Thus we shall assume that $A_{12}{}^1 + A_{22}{}^2 \neq 0$. It is not difficult to show that up to the action of the $ax + b$ group, if T is nilpotent, invariant, and non-trivial that either $T = \partial_{x^1} \otimes dx^2$ or $T = \partial_{x^2} \otimes dx^1$. We present the following examples to illustrate the nature of the invariants β_1 and β_2. We impose the relations of Theorem 15.4 to ensure the example are Bach flat in Case 1 and Case 2; we interchange the roles of x^1 and x^2 in Case 3 and Case 4 in Theorem 15.4.

Case 1. Let $\mathcal{M}$ be a Type $\mathcal{B}$ geometry with $\Gamma_{ij}{}^k = \frac{1}{x^1} A_{ij}{}^k$. Suppose that

$$A_{11}{}^2 = 0, \quad A_{11}{}^1 = 1, \quad T = \partial_{x^1} \otimes dx^2, \quad \text{and } \rho^{\tilde{\mathfrak{H}}} \text{ is non-degenerate}.$$

By Theorem 15.4, $\mathcal{N} = (T^*M, g_{\nabla,\Phi,T})$ is Bach flat. We have $\beta_1 = (A_{12}{}^1 + A_{22}{}^2)^2 \Delta^{-1}$ where

$$\begin{aligned}
\Delta = {} & 2(2 - A_{12}{}^2) A_{12}{}^2 (x^1)^2 \Phi_{11} - 4(2 - A_{12}{}^2)^2 A_{12}{}^2 x^1 x_3 \\
& -(4A_{12}{}^2 + 1)(A_{12}{}^1)^2 + 4(A_{12}{}^2 - 2) A_{22}{}^1 (A_{12}{}^2)^2 \\
& -(A_{22}{}^2)^2 + 2(1 - 2(A_{12}{}^2 - 1) A_{12}{}^2) A_{12}{}^1 A_{22}{}^2.
\end{aligned}$$

It now follows that $\beta_1 = 0$ if and only if the Ricci tensor ρ^∇ of $\mathcal{M}$ is symmetric. Moreover β_1 is a non-zero constant if and only if either $A_{12}{}^2 = 0$, in which case $\beta_1 = -\frac{(A_{12}{}^1 + A_{22}{}^2)^2}{(A_{12}{}^1 - A_{22}{}^2)^2}$, or $A_{12}{}^2 = 2$, and then $\beta_1 = -\frac{(A_{12}{}^1 + A_{22}{}^2)^2}{(3A_{12}{}^1 + A_{22}{}^2)^2}$. Further, if β_1 is non-zero, then β_2 is generically non-constant since

$$\begin{aligned}
\beta_2 = {} & \{(A_{12}{}^2 + 3)^2 (x^1)^2 \Phi_{11} + 2(A_{12}{}^2 - 2)(A_{12}{}^2 + 3)^2 x^1 x_3 \\
& -2(A_{12}{}^2 + 3)^2 A_{12}{}^2 A_{22}{}^1 - 2((A_{12}{}^2 - 1) A_{12}{}^2 + 3)(A_{22}{}^2)^2 \\
& -2((4A_{12}{}^2 + 9) A_{12}{}^2 + 6)(A_{12}{}^1)^2 \\
& -2((3A_{12}{}^2 - 4) A_{12}{}^2 - 9) A_{12}{}^1 A_{22}{}^2\} \Delta^{-1}.
\end{aligned}$$

Case 2. Let $\mathcal{M}$ be a Type $\mathcal{B}$ geometry with $\Gamma_{ij}{}^k = \frac{1}{x^1}A_{ij}{}^k$. Suppose

$$T = \partial_{x^1} \otimes dx^2, \quad A_{11}{}^2 = 0, \quad A_{12}{}^2 = A_{11}{}^1, \quad \text{and } \rho^{\tilde{\mathfrak{H}}} \text{ is non-degenerate}.$$

By Theorem 15.4, $\mathcal{N} = (T^*M, g_{\nabla,\Phi,T})$ is Bach flat. We have $\beta_1 = (A_{12}{}^1 + A_{22}{}^2)^2 \Delta^{-1}$ where

$$\begin{aligned}
\Delta \;=\;& 2A_{11}{}^1 (x^1)^2 \Phi_{11} - 4(A_{11}{}^1)^2 x^1 x_3 - (A_{22}{}^2)^2 \\
& -(4(A_{11}{}^1)^2 + 1)(A_{12}{}^1)^2 - 4A_{11}{}^1 A_{22}{}^1 + 2A_{12}{}^1 A_{22}{}^2 .
\end{aligned}$$

Therefore $\beta_1 = 0$ if and only if the Ricci tensor of $\mathcal{M}$ is symmetric. Moreover, one has that β_1 is a non-zero constant if and only if $A_{11}{}^1 = 0$, in which case $\beta_1 = -\frac{(A_{12}{}^1 + A_{22}{}^2)^2}{(A_{12}{}^1 - A_{22}{}^2)^2}$. Furthermore, if $\beta_1 \neq 0$, then

$$\begin{aligned}
\beta_2 \;=\;& \{ 4(A_{11}{}^1 + 1)^2 (x^1)^2 \Phi_{11} - 8(A_{11}{}^1 + 1)^2 A_{11}{}^1 x^1 x_3 \\
& -2(A_{11}{}^1 + 2)(A_{22}{}^2)^2 - 8(A_{11}{}^1 + 1)^2 A_{22}{}^1 \\
& -2(A_{11}{}^1(8A_{11}{}^1 + 9) + 2)(A_{12}{}^1)^2 + 4(3A_{11}{}^1 + 2)A_{12}{}^1 A_{22}{}^2 \} \Delta^{-1} .
\end{aligned}$$

Case 3. Let $\mathcal{M}$ be a Type $\mathcal{B}$ geometry with $\Gamma_{ij}{}^k = \frac{1}{x^1}A_{ij}{}^k$. Suppose

$$T = \partial_{x^2} \otimes dx^1, \quad A_{22}{}^1 = 0, \quad A_{22}{}^2 = 0, \quad \text{and } \rho^{\tilde{\mathfrak{H}}} \text{ is non-degenerate}.$$

We interchange the roles of x^1 and x^2 in Theorem 15.4 to see that $\mathcal{N} = (T^*M, g_{\nabla,\Phi,T})$ is Bach flat. We have $\beta_1 = (A_{12}{}^1)^2 \Delta^{-1}$ where

$$\begin{aligned}
\Delta \;=\;& (A_{12}{}^1)^2 \{ -2(x^1)^2 \Phi_{22} - 4A_{12}{}^1 x^1 x_4 \\
& -4A_{11}{}^1 A_{12}{}^2 + 4A_{11}{}^2 A_{12}{}^1 - 1 \} .
\end{aligned}$$

The invariant β_1 is never constant in this case. Moreover, if $\beta_1 \neq 0$, then

$$\begin{aligned}
\beta_2 \;=\;& (A_{12}{}^1)^2 \{ (x^1)^2 \Phi_{22} + 2A_{12}{}^1 x^1 x_4 - 12A_{12}{}^2 - 2A_{11}{}^2 A_{12}{}^1 \\
& -4 - 2(A_{11}{}^1)^2 - 8(A_{12}{}^2)^2 - 6(A_{12}{}^2 + 1)A_{11}{}^1 \} \Delta^{-1} .
\end{aligned}$$

It follows that β_2 is constant if and only if $2A_{11}{}^1 + 4A_{12}{}^2 + 3 = 0$, in which case $\beta_2 = -\frac{1}{2}$.

Case 4. Let $\mathcal{M}$ be a Type $\mathcal{B}$ geometry with $\Gamma_{ij}{}^k = \frac{1}{x^1}A_{ij}{}^k$. Suppose

$$T = \partial_{x^2} \otimes dx^1, \quad A_{22}{}^1 = 0, \quad A_{22}{}^2 = A_{12}{}^1, \quad \text{and } A_{12}{}^1 A_{12}{}^2 \neq 0.$$

We interchange the roles of x^1 and x^2 in Theorem 15.4 to see that $\mathcal{N} = (T^*M, g_{\nabla,\Phi,T})$ is Bach flat. The condition $A_{12}{}^1 A_{12}{}^2 \neq 0$ implies $\rho^{\tilde{\mathfrak{H}}}$ is non-degenerate. We then have

$$\begin{aligned}
\beta_1 \;&=\; -(A_{12}{}^2)^{-2} , \\
\beta_2 \;&=\; -\left((x^1)^2 \Phi_{22} - 2A_{12}{}^1 x^1 x_4 - 4(A_{12}{}^2)^2 - 2A_{12}{}^2 \right) (A_{12}{}^2)^{-2} .
\end{aligned}$$

In contrast with the previous cases, β_1 is constant while β_2 is never constant.

15.3   SIGNATURE $(2, 2)$ BACH FLAT MANIFOLDS

The material in this section arises from work of Calviño-Louzao et al. [14]. Since we are interested in constructing Bach flat manifolds which are not half-conformally-flat, we shall take T nilpotent in considering the metrics $g_{\nabla,\Phi,T}$. In Theorem 15.4, given T nilpotent, we showed how to construct an affine connection on the underlying manifold M so that $g_{\nabla,\Phi,T}$ was Bach flat. It follows from Theorem 15.14 that these geometries are VSI and we constructed scalar invariants which were not of Weyl type. In Section 15.2, we computed these invariants for various Type $\mathcal{B}$ geometries where T was nilpotent and where by Theorem 15.4, the resulting structures were Bach flat. In this section, we reverse the process and consider a perhaps more natural problem. We assume the affine structure $\mathcal{M}$ is given and look for non-trivial nilpotent operators so that $g_{\nabla,\Phi,T}$ is Bach flat. Since we are regarding the underlying structure Γ as fixed and looking for T, we cannot use Lemma 15.2 to renormalize the coordinate system. Our analysis is local. Since either $T^1{}_2(P) \neq 0$ or $T^2{}_1(P) \neq 0$, we assume for the sake of definiteness that $T^1{}_2(P) \neq 0$ and expand T near P in the form

$$T = \alpha(x^1, x^2) \begin{pmatrix} \xi(x^1, x^2) & 1 \\ -\xi^2(x^1, x^2) & -\xi(x^1, x^2) \end{pmatrix}. \tag{15.3.a}$$

Definition 15.17   We introduce the following operators:

$$\mathcal{P}_1(\xi) := -\xi^{(1,0)} + \xi\,\xi^{(0,1)} + \Gamma_{22}{}^1\xi^3 - (2\Gamma_{12}{}^1 - \Gamma_{22}{}^2)\xi^2 + (\Gamma_{11}{}^1 - 2\Gamma_{12}{}^2)\xi + \Gamma_{11}{}^2\,,$$

$$\begin{aligned}
\mathcal{P}_2(\xi,\alpha) := {}& \alpha\alpha^{(2,0)} + \xi^2\alpha\alpha^{(0,2)} - 2\xi\alpha\alpha^{(1,1)} + (\alpha^{(1,0)})^2 + \xi^2(\alpha^{(0,1)})^2 - 2\xi\alpha^{(1,0)}\alpha^{(0,1)} \\
& -\alpha\alpha^{(1,0)}\left(2\xi^{(0,1)} - 5\Gamma_{22}{}^1\xi^2 + 2(4\Gamma_{12}{}^1 - \Gamma_{22}{}^2)\xi - 3\Gamma_{11}{}^1 + 2\Gamma_{12}{}^2\right) \\
& +\alpha\alpha^{(0,1)}\left(2\xi\xi^{(0,1)} - 6\Gamma_{22}{}^1\xi^3 + (10\Gamma_{12}{}^1 - 3\Gamma_{22}{}^2)\xi^2 - 4(\Gamma_{11}{}^1 - \Gamma_{12}{}^2)\xi - \Gamma_{11}{}^2\right) \\
& +6\xi^4\alpha^2(\Gamma_{22}{}^1)^2 - 2\xi^3\alpha^2\left((\Gamma_{22}{}^1)^{(0,1)} + 9\Gamma_{12}{}^1\Gamma_{22}{}^1 - 3\Gamma_{22}{}^1\Gamma_{22}{}^2\right) \\
& -\xi^2\alpha^2\left(4\Gamma_{22}{}^1\xi^{(0,1)} - 3(\Gamma_{12}{}^1)^{(0,1)} - 2(\Gamma_{22}{}^1)^{(1,0)} + (\Gamma_{22}{}^2)^{(0,1)}\right. \\
& \qquad \left. - 12(\Gamma_{12}{}^1)^2 - (\Gamma_{22}{}^2)^2 - 7\Gamma_{11}{}^1\Gamma_{22}{}^1 + 7\Gamma_{12}{}^1\Gamma_{22}{}^2 + 9\Gamma_{12}{}^2\Gamma_{22}{}^1\right) \\
& +\xi\alpha^2\left(2(3\Gamma_{12}{}^1 - \Gamma_{22}{}^2)\xi^{(0,1)} - (\Gamma_{11}{}^1)^{(0,1)} - 3(\Gamma_{12}{}^1)^{(1,0)} + (\Gamma_{12}{}^2)^{(0,1)}\right. \\
& \qquad \left. +(\Gamma_{22}{}^2)^{(1,0)} - 2(\Gamma_{11}{}^1 - \Gamma_{12}{}^2)(4\Gamma_{12}{}^1 - \Gamma_{22}{}^2) + 4\Gamma_{11}{}^2\Gamma_{22}{}^1\right) \\
& -\alpha^2\left(2(\Gamma_{11}{}^1 - \Gamma_{12}{}^2)\xi^{(0,1)} - (\Gamma_{11}{}^1)^{(1,0)} + (\Gamma_{12}{}^2)^{(1,0)}\right. \\
& \qquad \left. -(\Gamma_{11}{}^1)^2 + \Gamma_{11}{}^1\Gamma_{12}{}^2 + 3\Gamma_{11}{}^2\Gamma_{12}{}^1 - \Gamma_{11}{}^2\Gamma_{22}{}^2\right)
\end{aligned}$$

Theorem 15.18   *Let (M, ∇) be an affine surface. Let T have the form given in Equation (15.3.a) and let Φ be arbitrary. The modified Riemannian extension $(T^*M, g_{\nabla,\Phi,T})$ of Equation (15.2.a) is Bach flat if and only if α and ξ are solutions to the partial differential equations $\mathcal{P}_1(\xi) = 0$ and $\mathcal{P}_2(\xi,\alpha) = 0$.*

Remark 15.19   Suppose $\mathcal{M}$ is real analytic. The operator $\mathcal{P}_1(\xi)$ of Definition 15.17 takes the form: $\mathcal{P}_1(\xi) = -\xi^{(1,0)} + \xi\xi^{(0,1)} + f(\xi, \Gamma)$. Given a real analytic function $\xi_0(x^2)$, the Cauchy–Kovalevski Theorem shows that there is a unique local solution to the equation $\mathcal{P}_1(\xi) = 0$ with $\xi(0, x^2) = \xi_0(x^2)$. Once ξ is determined, the operator $\mathcal{P}_2(\xi, \alpha)$ of Definition 15.17 takes the form $\mathcal{P}_2(\xi, \alpha) = \alpha\alpha^{(2,0)} - 2\xi\alpha\alpha^{(1,1)} + \xi^2\alpha\alpha^{(0,2)} + F(\alpha, d\alpha; \Gamma, d\Gamma; \xi, d\xi)$. Given real analytic functions $\alpha_0(x^2)$ and $\alpha_1(x^2)$, there exists a unique local solution to the equation $\mathcal{P}_2(\xi, \alpha) = 0$ with $\alpha(0, x^2) = \alpha_0(x^2)$ and $\alpha^{(1,0)}(0, x^2) = \alpha_1(x^2)$. Thus given ∇, there are many nilpotent T so that $\mathcal{N}$ is Bach flat in this setting; the auxiliary tensor Φ plays no role in the analysis.

A. Cauchy (1789–1857)   S. Kovalevskaya (1850–1891)

Proof. We suppose T is a nilpotent tensor field of Type $(1, 1)$. We then have that $\mathrm{Tr}\{T\} = 0$ and $\det(T) = 0$. If we assume that $T^1{}_2(P) \neq 0$, then T has the form given in Equation (15.3.a). A direct computation shows $\mathfrak{B}(\partial_{x^k}, \partial_{y_j}) = 0$ and $\mathfrak{B}(\partial_{y_i}, \partial_{y_j}) = 0$, and thus only $\mathfrak{B}_{11}$, $\mathfrak{B}_{12}$, and $\mathfrak{B}_{22}$, where $\mathfrak{B}_{ij} = \mathfrak{B}(\partial_{x^i}, \partial_{x^j})$, are relevant. We observe that

$$\text{Coefficient}[\mathfrak{B}_{11}, \alpha^{(2,0)}] = -4\alpha\xi^2, \quad \text{Coefficient}[\mathfrak{B}_{12}, \alpha^{(2,0)}] = -4\alpha\xi,$$
$$\text{Coefficient}[\mathfrak{B}_{22}, \alpha^{(2,0)}] = -4\alpha.$$

We therefore define $\mathfrak{Q}_1 := \mathfrak{B}_{11} - \mathfrak{B}_{12}\xi$, $\mathfrak{Q}_2 := \mathfrak{B}_{11} - \mathfrak{B}_{22}\xi^2$, and $\mathfrak{Q}_3 := 2\mathfrak{Q}_1 - \mathfrak{Q}_2$. We may then express $\mathfrak{Q}_3 = -4\alpha^2(\mathcal{P}_1)^2$ and thus the vanishing of $\mathfrak{Q}_3$ is equivalent to the vanishing of $\mathcal{P}_1$. We set $\mathcal{P}_1 = 0$ and express $\xi^{(1,0)} = F_{(1,0)}(\xi, \Gamma, \xi^{(0,1)})$. Differentiating this relation permits us to express

$$\xi^{(1,1)} = F_{(1,1)}(\xi, \Gamma, d\Gamma, \xi^{(0,1)}, \xi^{(0,2)}) \quad \text{and} \quad \xi^{(2,0)} = F_{(2,0)}(\xi, \Gamma, d\Gamma, \xi^{(0,1)}, \xi^{(0,2)}).$$

Substituting these relations then yields $\mathfrak{Q}_1 = 0$ and $\mathfrak{Q}_2 = 0$. Thus only $\mathfrak{B}_{11}$ plays a role. Substituting these relations permits us to complete the proof by expressing $\mathfrak{B}_{11} = -4\xi^2\mathcal{P}_2$, $\mathfrak{B}_{12} = -4\xi\mathcal{P}_2$, and $\mathfrak{B}_{22} = -4\mathcal{P}_2$. □

Bibliography

[1] Z. Afifi, Riemann extensions of affine connected spaces, *Quart. J. Math., Oxford Ser. (2)* **5** (1954), 312–320. DOI: 10.1093/qmath/5.1.312. 119

[2] T. Arias-Marco and O. Kowalski, Classification of locally homogeneous affine connections with arbitrary torsion on 2-dimensional manifolds, *Monatsh. Math.* **153** (2008), no. 1, 1–18. DOI: 10.1007/s00605-007-0494-0. 2, 27, 38, 39

[3] H. W. Brinkmann, Riemann spaces conformal to Einstein spaces, *Math. Ann.* **91** (1924), no. 3-4, 269–278. DOI: 10.1007/bf01556083. 123

[4] S. Bromberg and A. Medina, A note on the completeness of homogeneous quadratic vector fields on the plane, *Qual. Theory Dyn. Syst.* **6** (2005), no. 2, 181–185. DOI: 10.1007/bf02972671. 68

[5] M. Brozos-Vázquez and E. García-Río, Four-dimensional neutral signature self-dual gradient Ricci solitons, *Indiana Univ. Math. J.* **65** (2016), no. 6, 1921–1943. DOI: 10.1512/iumj.2016.65.5938. 13

[6] M. Brozos-Vázquez, E. García-Río, and P. Gilkey, Homogeneous affine surfaces: moduli spaces, *J. Math. Anal. Appl.* **444** (2016), no. 2, 1155–1184. DOI: 10.1016/j.jmaa.2016.07.005. 43, 59, 74

[7] M. Brozos-Vázquez, E. García-Río, and P. Gilkey, Homogeneous affine surfaces: affine Killing vector fields and gradient Ricci solitons, *J. Math. Soc. Japan* **70** (2018), no. 1, 25–70. DOI: 10.2969/jmsj/07017479. 43, 47, 56, 94, 99

[8] M. Brozos-Vázquez, E. García-Río, P. Gilkey, S. Nikčević, and R. Vázquez-Lorenzo, *The geometry of Walker manifolds*. Synthesis Lectures on Mathematics and Statistics **5**, Morgan & Claypool Publishers, Williston, VT, 2009. DOI: 10.2200/s00197ed1v01y200906mas005. 118

[9] M. Brozos-Vázquez, E. García-Río, P. Gilkey, and X. Valle-Regueiro, Solutions to the affine quasi-Einstein equation for homogeneous surfaces. https://arxiv.org/abs/1802.07953. 43, 80, 82

[10] M. Brozos-Vázquez, E. García-Río, P. Gilkey, and X. Valle-Regueiro, The affine quasi-Einstein equation for homogeneous surfaces, *Manuscripta Math.* **157** (2018), no. 1-2, 279–294. DOI: 10.1007/s00229-017-0987-7.

[11] M. Brozos-Vázquez, E. García-Río, P. Gilkey, and X. Valle-Regueiro, Half conformally flat generalized quasi-Einstein manifolds of metric signature $(2, 2)$, *Int. J. Math.* **29** (2018), no. 1, 1850002, 25 pp. DOI: 10.1142/s0129167x18500027. 124

[12] M. Brozos-Vázquez, E. García-Río, P. Gilkey, and X. Valle-Regueiro, A natural linear equation in affine geometry: the affine quasi-Einstein equation, *Proc. Amer. Math. Soc.* **146** (2018), no. 8, 3485–3497. DOI: 10.1090/proc/14090. 15, 26, 43

[13] E. Calviño-Louzao, E. García-Río, P. Gilkey, I. Gutiérrez-Rodríguez, and R. Vázquez-Lorenzo, Affine surfaces which are Kähler, para-Kähler, or nilpotent Kähler, *Results Math.* **73** (2018), no. 4, Art. 135, 24 pp. DOI: 10.1007/s00025-018-0895-5. 19, 102

[14] E. Calviño-Louzao, E. García-Río, P. Gilkey, I. Gutiérrez-Rodríguez, and R. Vázquez-Lorenzo, Constructing Bach flat manifolds of signature (2,2) using the modified Riemannian extension, *J. Math. Phys.* **60** (2019), no. 1, 013511, 14 pp. DOI: 10.1063/1.5080319. 117, 134

[15] H.-D. Cao and Q. Chen, On locally conformally flat gradient steady Ricci solitons, *Trans. Amer. Math. Soc.* **364** (2012), no. 5, 2377–2391.
DOI: 10.1090/s0002-9947-2011-05446-2. 123

[16] F. Catoni, R. Cannata, V. Catoni, and P. Zampetti, Lorentz surfaces with constant curvature and their physical interpretation, *Nuovo Cimento Soc. Ital. Fis. B* **120** (2005), no. 1, 37–51. DOI: 10.1393/ncb/i2004-10129-3. 111

[17] M. Christ, Some nonanalytic-hypoelliptic sums of squares of vector fields, *Bull. Amer. Math. Soc. (N.S.)* **26** (1992), no. 1, 137–140.
DOI: 10.1090/s0273-0979-1992-00258-6. 14

[18] A. Coley, S. Hervik, D. McNutt, N. Musoke, and D. Brooks, Neutral signature Walker-VSI metrics, *Class. Quantum Grav.* **31** (2014), 035015 (14pp). DOI: 10.1088/0264-9381/31/3/035015. 125

[19] V. Cortés, C. Mayer, T. Mohaupt, and F. Saueressig, Special geometry of Euclidean supersymmetry. I. Vector multiplets, *J. High Energy Phys.* (2004), no. 3, 028, 73 pp. DOI: 10.1088/1126-6708/2004/03/028. 20

[20] V. Cruceanu, P. Fortuny, and P. M. Gadea, A survey on paracomplex geometry, *Rocky Mountain J. Math.* **26** (1996), no. 1, 83–115. DOI: 10.1216/rmjm/1181072105. 111

[21] D. D'Ascanio, P. Gilkey, and P. Pisani, Geodesic completeness for type $\mathcal{A}$ surfaces, *Differential Geom. Appl.* **54** (2017), part A, 31–43. DOI: 10.1016/j.difgeo.2016.12.008. 65

[22] D. D'Ascanio, P. Gilkey, and P. Pisani, The geometry of locally symmetric affine surfaces. *Vietnam J. Math* **47** (2019), 5–21. DOI: 10.1007/s10013-018-0280-4. 29, 112, 113

[23] D. D'Ascanio, P. Gilkey, and P. Pisani, Affine symmetric spaces with torsion, *in preparation*. 2, 9, 10, 104

[24] D. D'Ascanio, P. Gilkey, and P. Pisani, Affine Killing vector fields on homogeneous surfaces with torsion, *in preparation*. 39

[25] A. Derdzinski, Connections with skew-symmetric Ricci tensor on surfaces, *Results Math.* **52** (2008), no. 3-4, 223–245. DOI: 10.1007/s00025-008-0307-3. 16

[26] A. Derdzinski and W. Roter, Walker's theorem without coordinates, *J. Math. Phys.* **47** (2006), no. 6, 062504, 8 pp. DOI: 10.1063/1.2209167. 119

[27] S. Dumitrescu, Locally homogeneous rigid geometric structures on surfaces, *Geom. Dedicata* **160** (2012), 71–90. DOI: 10.1007/s10711-011-9670-4.

[28] S. Dumitrescu and A. Guillot, Quasihomogeneous analytic affine connections on surfaces, *J. Topol. Anal.* **5** (2013), no. 4, 491–532. DOI: 10.1142/s1793525313500222.

[29] L. P. Eisenhart, *Non-Riemannian geometry*. American Mathematical Society Colloquium Publications **8**, American Mathematical Society, Providence, RI, 1990. DOI: 10.1090/coll/008. 11

[30] P. Gilkey, *Invariance theory, the heat equation, and the Atiyah–Singer index theorem*. Second edition. Studies in Advanced Mathematics, CRC Press, Boca Raton, FL, 1995. 14

[31] P. Gilkey, The moduli space of type A surfaces with torsion and non-singular symmetric Ricci tensor, *J. Geom. Phys.* **110** (2016), 69–77. DOI: 10.1016/j.geomphys.2016.07.012. 2

[32] P. Gilkey, Moduli spaces of type B surfaces with torsion, *J. Geom.* **108** (2017), no. 2, 637–653. DOI: 10.1007/s00022-016-0364-9. 2

[33] P. Gilkey and S. Nikčević, Complete k-curvature homogeneous pseudo-Riemannian manifolds, *Ann. Global Anal. Geom.* **27** (2005), no. 1, 87–100. DOI: 10.1007/s10455-005-5217-y. 125, 128

[34] P. Gilkey, J. H. Park, and X. Valle-Regueiro, Affine Killing complete and geodesically complete homogeneous affine surfaces, *J. Math. Anal. Appl.* **474** (2019), no. 1, 179–193. DOI: 10.1016/j.jmaa.2019.01.038. 65

[35] P. Gilkey and X. Valle-Regueiro, Applications of PDEs to the study of affine surface geometry, *Matematicki Vesnik* **71** (2019), 45–62. 24, 43, 47, 59

[36] A. R. Gover and P. Nurowski, Obstructions to conformally Einstein metrics in n dimensions, *J. Geom. Phys.* **56** (2006), no. 3, 450–484. DOI: 10.1016/j.geomphys.2005.03.001. 123

[37] A. Guillot and A. Sánchez Godinez, A classification of locally homogeneous affine connections on compact surfaces, *Ann. Global Anal. Geom.* **46** (2014), no. 4, 335–349. DOI: 10.1007/s10455-014-9426-0. 28

[38] O. Kobayashi, A differential equation arising from scalar curvature function, *J. Math. Soc. Japan* **34** (1982), no. 4, 665–675. DOI: 10.2969/jmsj/03440665. 123

[39] S. Kobayashi and K. Nomizu, *Foundations of differential geometry. Vol. I and Vol. II.* Wiley Classics Library. A Wiley-Interscience Publication. John Wiley & Sons, Inc., New York, 1996. 2, 3, 11, 15

[40] O. Kobayashi and M. Obata, Certain mathematical problems on static models in general relativity. *Proc. of the 1980 Beijing Symposium on Differential Geometry and Differential Equations, Vol. 1, 2, 3,* (Beijing, 1980), 1333–1343, Science Press Beijing, Beijing, 1982. 123

[41] S. S. Koh, On affine symmetric spaces, *Trans. Amer. Math. Soc.* **119** (1965), 291–309. DOI: 10.1090/s0002-9947-1965-0184170-2. 9

[42] O. Kowalski, B. Opozda, and Z. Vlášek, A classification of locally homogeneous affine connections with skew-symmetric Ricci tensor on 2-dimensional manifolds, *Monatsh. Math.* **130** (2000), no. 2, 109–125. DOI: 10.1007/s006050070041. 28

[43] O. Kowalski, B. Opozda, and Z. Vlášek, On locally nonhomogeneous pseudo-Riemannian manifolds with locally homogeneous Levi-Civita connections, *Int. J. Math.* **14** (2003), no. 5, 559–572. DOI: 10.1142/s0129167x03001971. 29, 39

[44] O. Kowalski, B. Opozda, and Z. Vlášek, A classification of locally homogeneous connections on 2-dimensional manifolds via group-theoretical approach, *Cent. Eur. J. Math.* **2** (2004), no. 1, 87–102. DOI: 10.2478/bf02475953. 27

[45] O. Munteanu and N. Sesum, On gradient Ricci solitons, *J. Geom. Anal.* **23** (2013), no. 2, 539–561. DOI: 10.1007/s12220-011-9252-6. 123

[46] A. Newlander and L. Nirenberg, Complex analytic coordinates in almost complex manifolds, *Ann. of Math. (2)* **65** (1957), 391–404. DOI: 10.2307/1970051. 20

[47] K. Nomizu, Invariant affine connections on homogeneous spaces, *Amer. J. Math.* **76** (1954), 33–65. DOI: 10.2307/2372398. 9

[48] K. Nomizu, On local and global existence of Killing vector fields, *Ann. of Math. (2)* **72** (1960), 105–120. DOI: 10.2307/1970148. 53

[49] P. J. Olver, *Equivalence, Invariants, and Symmetry.* Cambridge University Press, Cambridge, 1995. DOI: 10.1017/cbo9780511609565. 28

[50] B. Opozda, Locally symmetric connections on surfaces, *Results Math.* **20** (1991), no. 3-4, 725–743. DOI: 10.1007/bf03323207. 9, 65

[51] B. Opozda, A class of projectively flat surfaces, *Math. Z.* **219** (1995), no. 1, 77–92. DOI: 10.1007/bf02572351. 10

[52] B. Opozda, Affine versions of Singer's theorem on locally homogeneous spaces, *Ann. Global Anal. Geom.* **15** (1997), no. 2, 187–199. DOI: 10.1023/A:1006585424144. 9

[53] B. Opozda, A classification of locally homogeneous connections on 2-dimensional manifolds, *Differential Geom. Appl.* **21** (2004), no. 2, 173–198. DOI: 10.1016/j.difgeo.2004.03.005. xiii, 27, 38

[54] B. Opozda, Locally homogeneous affine connections on compact surfaces, *Proc. Amer. Math. Soc.* **132** (2004), no. 9, 2713–2721. DOI: 10.1090/S0002-9939-04-07402-7. 28

[55] J. Patera, R. T. Sharp, P. Winternitz, and H. Zassenhaus, Invariants of real low dimension Lie algebras, *J. Math. Phys.* **17** (1976), no. 6, 986–994. DOI: 10.1063/1.522992. 34, 56

[56] E. M. Patterson and A. G. Walker, Riemann extensions, *Quart. J. Math., Oxford Ser. (2)* **3** (1952), 19–28. DOI: 10.1093/qmath/3.1.19. 119

[57] V. Pecastaing, On two theorems about local automorphisms of geometric structures, *Ann. Inst. Fourier (Grenoble)* **66** (2016), no 1, 175–208. DOI: 10.5802/aif.3009. 9

[58] C. Steglich, Invariants of conformal and projective structures, *Results Math.* **27** (1995), no. 1-2, 188–193. DOI: 10.1007/bf03322280. 11

[59] F. Trèves, Analytic-hypoelliptic partial differential equations of principal type, *Comm. Pure Appl. Math.* **24** (1971), 537–570. DOI: 10.1002/cpa.3160240407. 14

[60] A. Vanžurová, On metrizability of locally homogeneous affine 2-dimensional manifolds, *Arch. Math. (Brno)* **49** (2013), no. 5, 347–357. DOI: 10.5817/am2013-5-347. 39

[61] A. G. Walker, Canonical form for a Riemannian space with a parallel field of null planes, *Quart. J. Math., Oxford Ser. (2)* **1** (1950), 69–79. DOI: 10.1093/qmath/1.1.69. 118, 119

[62] Y.-C. Wong, Two dimensional linear connexions with zero torsion and recurrent curvature, *Monatsh. Math.* **68** (1964), 175–184. DOI: 10.1007/bf01307120. 15, 16, 17

Authors' Biographies

ESTEBAN CALVIÑO-LOUZAO[1] is a member of the research group in Riemannian Geometry at the Department of Geometry and Topology of the University of Santiago de Compostela (Spain). He received his Ph.D. in 2011 from the University of Santiago under the direction of E. García-Río and R. Vázquez-Lorenzo. His research specialty is Riemannian and pseudo-Riemannian geometry. He has published more than 20 research articles and books.

EDUARDO GARCÍA-RÍO[2] is a Professor of Mathematics at the University of Santiago de Compostela (Spain). He is a member of the editorial board of Differential Geometry and its Applications and The Journal of Geometric Analysis and leads the research group in Riemannian Geometry at the Department of Geometry and Topology of the University of Santiago de Compostela (Spain). He received his Ph.D. in 1992 from the University of Santiago under the direction of A. Bonome and L. Hervella. His research specialty is Differential Geometry. He has published more than 120 research articles and books.

[1] Dir. Xeral de Educación, Formación Profesional e Innovación Educativa, San Caetano, s/n, 15781 Santiago de Compostela, Spain.
email: estebcl@edu.xunta.es
[2] Department of Mathematics, Faculty of Mathematics, University of Santiago de Compostela, 15782 Santiago de Compostela, Spain
email: eduardo.garcia.rio@usc.es

PETER B. GILKEY[3] is a Professor of Mathematics and a member of the Institute of Theoretical Science at the University of Oregon. He is a fellow of the American Mathematical Society and is a member of the editorial board of Results in Mathematics, Differential Geometry and its Applications, and The Journal of Geometric Analysis. He received his Ph.D. in 1972 from Harvard University under the direction of L. Nirenberg. His research specialties are Differential Geometry, Elliptic Partial Differential Equations, and Algebraic topology. He has published more than 275 research articles and books.

JEONGHYEONG PARK[4] is a Professor of Mathematics at Sungkyunkwan University and is an associate member of the KIAS (Korea). She received her Ph.D. in 1990 from Kanazawa University in Japan under the direction of H. Kitahara. Her research specialties are spectral geometry of Riemannian submersion and geometric structures on manifolds like eta-Einstein manifolds and H-contact manifolds. She organized the geometry section of AMC 2013 (he Asian Mathematical Conference 2013), the ICM 2014 satellite conference on Geometric anal-ysis, and geometric structures on manifolds (2016). She has published more than 90 research papers and books.

[3]Mathematics Department, University of Oregon, Eugene OR 97403 U.S.
  email: gilkey@uoregon.edu
[4]Mathematics Department, Sungkyunkwan University, Suwon, 16419, Korea
  email: parkj@skku.edu

RAMÓN VÁZQUEZ-LORENZO[5] is a member of the research group in Riemannian Geom-etry at the Department of Geometry and Topology of the University of Santiago de Compostela (Spain). He is a member of the Spanish Research Network on Relativity and Gravitation. He received his Ph.D. in 1997 from the University of Santiago de Compostela under the direction of E. García-Río and R. Castro. His research focuses mainly on Differential Geometry, with special emphasis on the study of the curvature and the algebraic properties of curvature operators in the Lorentzian and in the higher signature settings. He has published more than 60 research articles and books.

[5]Department of Geometry and Topology, Faculty of Mathematics, University of Santiago de Compostela, 15782 Santiago de Compostela, Spain.
email: ravazlor@edu.xunta.es

Index

Printed in the United States
by Baker & Taylor Publisher Services